AF352855

# Synthesis and Characterization of Ferroelectrics

# Synthesis and Characterization of Ferroelectrics

Editor

**Jan Dec**

MDPI • Basel • Beijing • Wuhan • Barcelona • Belgrade • Manchester • Tokyo • Cluj • Tianjin

*Editor*
Jan Dec
University of Silesia
Poland

*Editorial Office*
MDPI
St. Alban-Anlage 66
4052 Basel, Switzerland

This is a reprint of articles from the Special Issue published online in the open access journal *Crystals* (ISSN 2073-4352) (available at: https://www.mdpi.com/journal/crystals/special_issues/characterization_ferroelectrics#).

For citation purposes, cite each article independently as indicated on the article page online and as indicated below:

LastName, A.A.; LastName, B.B.; LastName, C.C. Article Title. *Journal Name* **Year**, *Article Number*, Page Range.

**ISBN 978-3-03943-655-2 (Hbk)**
**ISBN 978-3-03943-656-9 (PDF)**

# Contents

# About the Editor

**Jan Dec** graduated from High Pedagogical School, Opole, Poland in physics in 1971, and received his Ph.D. in Physics from the Institute of Physics, University of Silesia, Katowice, Poland in 1980. After working as a post-doctoral fellow, Dr. Dec was hired as an Associate Professor in 1997 and then, in 2000, got a Professor position at the same university. Prof. Dec has extensive foreign scientific experience. He has worked at the University of Rostov-on-Don (The Soviet Union), the University of Duisburg (Germany), and the University of Oxford (United Kingdom), among others. His current field of interest focuses on low frequency dielectric relaxation in electroactive materials such as ferroelectrics, relaxor ferroelectrics, and ferroic glasses. According to the Web of Science, his H-index = 32.

# Preface to "Synthesis and Characterization of Ferroelectrics"

Nowadays, it is difficult to imagine efficient medical diagnostics without ultrasonography or mid-range car lacking sensors. Piezo lighters and motion sensors are in common use and a new generation of radars, without any moving parts, have successfully been developed. All these exemplary technical achievements were made possible by the discovery of ferroelectricity in Rochelle Salt 100 years ago by Joseph Valasek—an American physicist of Czech origin—who first observed the ferroelectric hysteresis loop. Subsequently, more and more new ferroelectric compounds were synthetized in the middle of the 20th century.

Ferroelectrics are usually considered as multifunctional materials. Hence, their miscellaneous applications, which vary from their generic property, emerge. In addition, ferroelectrics exhibit highly non-linear responses, which are changeable rather than fixed, mimicking, to a large extent, biological systems. Consequently, this kind of behavior is qualified as "smart" and respective compounds are termed as "smart materials".

Within the last 100 years, ferroelectric materials have been able to efficiently adjust to devices in our daily life (e.g., ultrasound or thermal imaging, gyroscopes, filters, accelerometers). Promising breakthrough applications, such as solid-state refrigeration, non-volatile memories, optical devices, energy harvesting, and energy storage appliances, are still under development, making ferroelectrics one of tomorrow's most important materials.

This Special Issue on "Synthesis and Characterization of Ferroelectrics" covers a broad range of physical properties of ferroelectrics, examines their technological aspects, and contains a mixture of review articles and original contributions.

**Jan Dec**
*Editor*

 *crystals*

*Editorial*

# Synthesis and Characterization of Ferroelectrics

**Jan Dec**

Institute of Materials Science, University of Silesia, Bankowa Str. 12, 40-007 Katowice, Poland;
jan.dec@us.edu.pl

Received: 16 September 2020; Accepted: 16 September 2020; Published: 17 September 2020

Ferroelectrics belong to one of the most studied groups of materials in terms of research and applications. Apart from their foremost property (the ferroelectricity), these materials also display other numerous attractive properties such as piezoelectricity, pyroelectricity, electrocaloric and electro-optic effects, etc., which designate them as multifunctional materials. Therefore, these materials are suitable for a wide range of applications ranging from effective sensors, transducers and actuators to optical and memory devices. Since the discovery of ferroelectricity in Rochelle salt in 1920 by Valasek [1], numerous applications using such effects have been developed. In addition, ferroelectrics, and other ferroics, exhibit a highly non-linear response, which is changeable rather than fixed, mimicking, to a large extent, biological systems [2]. Consequently, this kind of behavior is qualified as "smart" and respective systems are termed as "smart materials" [2]. This Special Issue on "Synthesis and Characterization of Ferroelectrics" covers a broad range of physical properties of ferroelectrics, their technological aspects and contains a mixture of review article and original contributions.

We start with the review paper by Liu et al. [3], which summarizes the recent progress on lead-free (BaCa)(ZrTi)O$_3$ (BCT-BZT for short) piezoelectric ceramics. Different substitution mechanisms offer some thoughts towards the future improvement of BCT-BZT ceramics including the electrocaloric effect, fluorescence and energy storage. Element segregation along axial and radial directions and electrical properties of a relaxor-based single crystal with nominal composition of 0.68Pb(Mg$_{1/3}$Nb$_{2/3}$)-0.32PbTiO$_3$ (PMN-32PT) were investigated by Wang et al. [4]. It is found that such segregation differently influences the electrical properties of the investigated system. While the electrical properties along the axial direction strongly depend on the PbTiO$_3$ content, the electrical properties along the radial direction are mainly determined by the ratio of Nb and Mg. Another technological route is presented by Li et al. [5]. The authors investigated dielectric and conductivity mechanisms of Fe-substituted PMN-32PT crystals. This heterovalent ionic substitution led to enhancement of the coercive field due to wall-pinning induced by charged defect dipoles. On the other hand, the dominating conduction carriers are electrons arising from the first ionization of oxygen vacancies.

Two papers are dedicated to thin-film capacitor applications. The results on structure and electrical properties of lead-free Na$_{0.5}$Bi$_{0.5}$TiO$_3$ based epitaxial films are reported by Song et al. [6]. Pt/Na$_{0.5}$Bi$_{0.5}$TiO$_3$/La$_{0.5}$Sr$_{0.5}$CoO$_3$ (Pt/NBT/LSCO) was fabricated on a (110) SrTiO$_3$ substrate. Both NBT and LSCO films displayed (110) epitaxial growth. The PT/NBT/LSCO capacitor possesses good fatigue resistance and retention, as well as ferroelectric properties. Another lead-free thin-film capacitor is based on Ba$_{0.3}$Sr$_{0.7}$Zr$_{0.18}$Ti$_{0.82}$O$_3$ (BSZT) compound [7]. The obtained BSZT films feature a low leakage current density of the order of $7.65 \times 10^{-7}$ A/cm$^2$, and breakdown strength as high as 4 MV/cm. In addition, these films exhibit an almost linear and acceptable temperature change of capacitance ($\Delta C/C \approx 13.6\%$) and also large capacitance density of the order of 1.7 nF/mm$^2$ at 100 kHz.

Finally, the Special Issue ends with a report on an enhanced electrocaloric effect, as observed by Lu et al. [8], in 0.73Pb(Mg$_{1/3}$Nb$_{2/3}$)O$_3$-0.27 PbTiO$_3$ single crystals. The authors claim that a directly measured change in temperature $\Delta T$ >2.5 K of the sample may be observed under an external electrical field which was reversed at room temperature from 1 MV/m to −1MV/m. The reported temperature change is larger than that deduced according to the Maxwell relation and larger than that calculated using the Landau–Ginsburg–Devonshire phenomenological theory. We hope that this contribution will stimulate further research for effective solid-state refrigeration materials as well as refreshing discussion concerned with the investigation methodology of the electrocaloric effect.

The present Special Issue on "Synthesis and Characterization of Ferroelectrics" can be considered as a status report reviewing some progress that has been achieved over the past few years in selected topic areas related to ferroelectric materials.

## References


1.   Valasek, J. Piezoelectric and Allied Phenomena in Rochelle Salt. *Phys. Rev.* **1920**, *15*, 537.
2.   Wadhavan, V.K.; Pandit, P.; Gupta, S.M. PMN-PT Based Relaxor Ferroelectrics as Very Smart Materials. *Materials Sci. Eng. B* **2005**, *120*, 199. [CrossRef]
3.   Liu, W.; Cheng, L.; Li, S. Prospective of (BaCa)(ZrTi)$O_3$ Lead-free Piezoelectric Ceramics. *Crystals* **2019**, *9*, 179. [CrossRef]
4.   Wang, S.; Xi, Z.; Fang, P.; Li, X.; Long, W.; He, A. Element Segregation and Electrical Properties of PMN-32PT Grown Using Bridgman Method. *Crystals* **2019**, *9*, 98. [CrossRef]
5.   Li, X.; Fan, X.; Xi, Z.; Liu, P.; Long, W.; Fang, P.; Guo, F.; Nan, R. Dielectric Relaxor and Conductivity Mechanis, in Fe-Substituted PMN-32PT Ferroelectric Crystal. *Crystals* **2019**, *9*, 241. [CrossRef]
6.   Song, J.; Gao, J.; Zhang, S.; Luo, L.; Dai, X.; Zhao, L.; Liu, B. Structure and Electrical Properties of $Na_{0.5}Bi_{0.5}TiO_3$ Epitaxial Films with (110) Orientation. *Crystals* **2019**, *9*, 558. [CrossRef]
7.   Chen, X.; Mo, T.; Huang, B.; Liu, Y.; Yu, P. Capacitance Properties in $Ba_{0.3}Sr_{0.7}Zr_{0.18}Ti_{0.82}O_3$ Thin Films on Silicon Substrate for Thin Film Capacitor Applications. *Crystals* **2020**, *10*, 318. [CrossRef]
8.   Lu, B.; Jian, X.; Lin, X.; Yao, Y.; Tao, T.; Liang, B.; Luo, H.; Lu, S.-G. Enhanced Electrocaloric Effect in $0.73Pb(Mg_{1/3}Nb_{2/3})O_3$-$0.27\,PbTiO_3$ Single Crystals via Direct Measurements. *Crystals* **2020**, *10*, 451. [CrossRef]

*Article*

# Element Segregation and Electrical Properties of PMN-32PT Grown Using the Bridgman Method

Sijia Wang [1,2], Zengzhe Xi [1,2,3,*], Pinyang Fang [2,3], Xiaojuan Li [2,3], Wei Long [2,3] and Aiguo He [2,3]

1   School of Science, Xi'an Technological University, Xi'an 710021, China; cigawang1231@163.com
2   Shaanxi Key Laboratory of Photoelectric Functional Materials and Devices, Xi'an 710021, China;
    fpy_2000@163.com (P.F.); lixiaojuan28@163.com (X.L.); longwei@xatu.com (W.L.);
    5129226701@163.com (A.H.)
3   School of Materials and Chemical Engineering, Xi'an Technological University, Xi'an 710021, China
*   Correspondence: zzhxi@xatu.edu.cn; Tel.: +86-29-86173324

Received: 5 January 2019; Accepted: 13 February 2019; Published: 15 February 2019

**Abstract:** A single crystal with nominal composition $Pb(Mg_{1/3}Nb_{2/3})O_3$-32$PbTiO_3$ (PMN-32PT) was grown by the Bridgman technique. Crystal orientation was determined using the rotating orientation X-ray diffraction (RO-XRD). Element distribution was measured along different directions using inductively coupled plasma-mass spectrometry (ICP-MS). The effect of the element segregation along axial and radial directions on the electrical properties of the PMN-32PT crystal was investigated. It is indicated that the electrical properties of the samples along the axial direction were strongly dependent on the PT ($PbTiO_3$) content. With the increase of the PT content, the piezoelectric coefficient and remnant polarization were improved. Differently, the electrical properties of the samples along the radial direction were mainly determined by the ratio of the Nb and Mg. The reasons for the element segregation and electrical properties varied with the composition were discussed.

**Keywords:** PMN-32PT; characterization; segregation; Bridgman technique; ferroelectric materials

---

## 1. Introduction

Relaxor-based ferroelectric single crystals $Pb(Mg_{1/3}Nb_{2/3})O_3$-$PbTiO_3$ (PMN-PT) have an ultrahigh piezoelectric coefficient ($d_{33} > 2500$ pC/N), an electromechanical coupling factor ($k_{33} > 0.95$), and a low dielectric loss compared to traditional piezoelectric ceramics [1–5]. Based on these superior properties, PMN-PT single crystals are usually considered promising materials in sensors, ultrasonic transducers, and motors applications [6–9].

Large sized PMN-PT single crystals are grown mainly using the Bridgman technique [10–12]. Based on this technique, researchers have further managed to improve the di/piezoelectric properties of the PMN-PT system through some effective ways [13–17]. Hu et al. [18] verified that the high-temperature poling technique was contributed to the enhanced piezoelectric properties. Recently, it was discovered that the optical properties could be induced by rare-earth ions doping in the PMN-PT system [19,20]. Xi et al. [21] confirmed that the specific absorption at the UV-VIS-NIR band and the strong green and red up-conversion photoluminescence (UC PL) under 980 nm laser excitation were observed in the $Er^{3+}$- and $Er^{3+}/Yb^{3+}$-modified $Pb(Sc_{1/2}Nb_{1/2})O_3$-$Pb(Mg_{1/3}Nb_{2/3})O_3$-$PbTiO_3$ (PSN-PMN-PT) crystals using the flux method. For the Bridgman technique, it is confirmed that the element segregation exists in the single crystals and the electrical properties of the single crystals are strongly dependent on the compositions [22–24]. These results indicate that the element segregation occurred during the growth of the PMN-PT single crystals, and the segregation of PT led to inhomogeneity in electrical properties along the axial direction. Unfortunately, the element segregation along the radial direction and the reasons for the Nb and Mg segregation of the PMN-PT single crystals using the Bridgman technique was rarely reported in the literature.

In this study, the single crystal with nominal compositional PMN-32PT was grown by the Bridgman technique. The element distribution along the axial and radial directions was confirmed by the inductively coupled plasma–mass spectrometry (ICP-MS). The effect of the element segregation along the axial and radial direction on the electrical properties of the PMN-32PT crystals was investigated. The reasons that the element segregation and electrical properties varied with the composition along the axial direction were also discussed.

## 2. Experimental Procedure

A PMN-32PT single crystal (Ø25 mm) was grown by the Bridgman method (Figure 1a). The crystal was faint yellow with good transparency. Some stress-induced cracks were presented on the top of the crystal. A sheet with a thickness of 0.8 mm was cut from the as-grown crystal boule along the axial direction (Figure 1b), which showed a poor uniformity in color. The sheet was divided along axial and radial directions with a size of 2 × 2 mm. The specimens were named as test points from Y1 to Y15 along the axial direction and test points from X1 to X4 along the radial direction. The Y9 and X4 test points were the same (shown in Figure 2).

(a)        (b)

**Figure 1.** The as-grown PMN-32PT single crystal and its axial section: (**a**) the as-grown PMN-32PT single crystal; (**b**) the axial section along the length of the crystal.

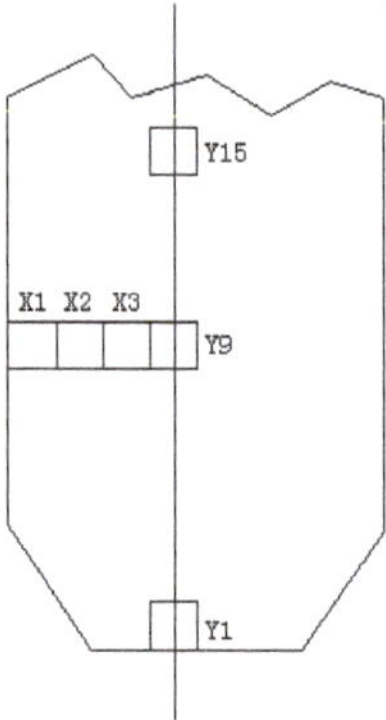


**Figure 2.** A sketch of specimen cutting.

The orientation of the single crystal was analyzed using rotating orientation X-ray diffraction (RO-XRD) (D/max-2550, Rigaku Corporation, Tokyo, Japan, 2004). Before the electrical property measurements, the obtained specimens were annealed at 300 °C for 1.5 h to eliminate stress. Silver paste was used to cover the two faces of the crystal sample that were used as electrodes. The dielectric properties were measured using an impedance analyzer (Agilent 4294A, Agilent Technologies Inc., Santa Clara, CA, USA, 2012) at 1 kHz at room temperature. The ferroelectric properties were measured using a ferroelectric analyzer (Radiant Precision PremierII, Radiant Technologies Inc., Albuquerque, NM, USA, 2005) at room temperature. For the piezoelectric constant ($d_{33}$) test, the specimens were poled at room temperature in silicone oil under an applied electric field of 1280 kV/mm for 15 min. The piezoelectric constant was measured using a quasistatic meter (ZJ-6A, Institute of Acoustics Academic Sinica, Beijing, China, 2005). The element analysis was performed using the ICP-MS (NexION 350D, PerkinElmer, Waltham, MA, USA, 2017) after the samples were dissolved in a mixture of concentrated nitric acid and hydrofluoric acid. In order to calibrate the errors of the quantitative analysis, blank experiments were used before the element analysis.

## 3. Results and Discussion

### 3.1. RO-XRD

The RO-XRD patterns of the PMN-PT sample are shown in Figure 3. Two strong peaks were observed at 22.88° and 35.09° and there were no other diffraction peaks between them, indicating that the sample is a single crystal. Based on the fixed angle φ between the two crystal planes, the oriented direction of one crystal plane can be determined by the other [25,26]. In this study, the (211), (220) and (222) crystal planes were selected to calculate the orientation of the samples according to the following equation [27,28]:

$$\varphi = \frac{\theta_2 - \theta_1}{2} \tag{1}$$

where $\theta_1$ and $\theta_2$ are the degrees of the strong diffraction peaks, respectively. The calculated results show the crystal plane perpendicular to the axial direction is (432). Along the axial direction, the crystal plane belongs to {771}.

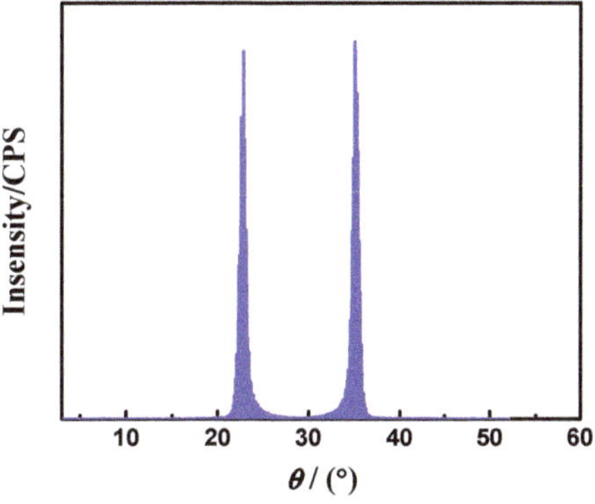

**Figure 3.** The RO-XRD pattern with 2θ = 56° of the PMN-32PT crystal.

### 3.2. Axis Distribution

The distribution of elements along the axial direction for the PMN-32PT single crystals is shown in Figure 4. Obviously, the Ti content exhibits an increasing trend along the axial direction from the bottom to the top. On the other hand, the content of the Nb and Mg decreases from the bottom to the top. The variations of the mass fraction of the Nb and Mg are calculated to be about 2.29% and 0.97%, respectively. The segregation during the growth of the PMN-PT single crystals is responsible for the variation of the elements [29].

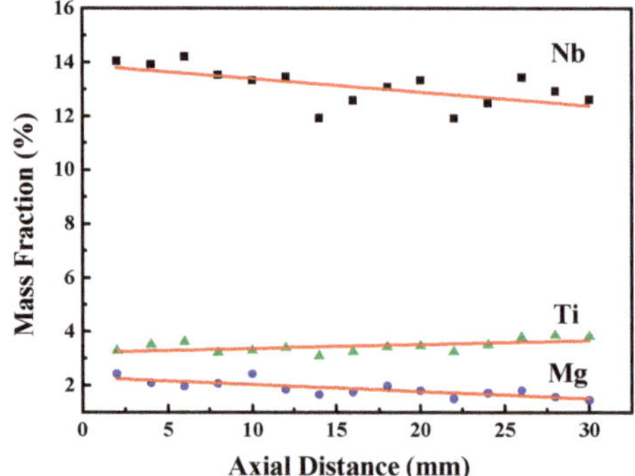

**Figure 4.** The composition distribution along the axis of the PMN-32PT: points represent the experimental data and solid lines represent the fitting.

Generally, the segregation of the element depends on its the effective segregation coefficient $k$ during the crystal growth process, which can be obtained from the following equation [30]:

$$C_s = kC_0(1 - f)^{(k-1)} \tag{2}$$

where $C_S$ and $C_0$ are the concentrations of the solid and initial melt, respectively, and $f$ is the fraction of the melt solidified corresponding to $C_S$. Normally, as the effective segregation coefficient is less than 1 ($k < 1$), the element content displays an increasing tendency along the axial direction from the bottom to the top. As the effective segregation coefficient is more than 1 ($k > 1$), the element content shows a decreasing tendency from the bottom to the top of the crystal. In this work, the effective segregation coefficients $k$ are 1.20, 1.03 and 0.82 for Mg, Nb and Ti, respectively. The $k$ for Ti is slightly lower than that obtained by Benayad [23] in a PMN-40PT system ($k = 0.849$) and Zawilski [24] in a PMN-35PT system ($k = 0.84$), which is attributed to the slower solidification rate of the present work. Table 1 provides the molar percentages of Mg, Nb and Ti along the axial direction of the crystal from the bottom to the top.

**Table 1.** The variation of composition from the bottom to the top.

| Distance(mm) | Mg (mol%) | Nb (mol%) | Ti (mol%) |
| --- | --- | --- | --- |
| 0–2 | 31.40 ± 0.16 | 47.11 ± 0.22 | 21.49 ± 0.38 |
| 2–4 | 28.31 ± 0.28 | 48.12 ± 0.58 | 23.57 ± 0.76 |
| 4–6 | 26.02 ± 0.14 | 49.44 ± 0.28 | 24.54 ± 0.42 |
| 6–8 | 28.69 ± 0.01 | 48.77 ± 0.04 | 22.55 ± 0.06 |
| 8–10 | 32.52 ± 0.24 | 45.53 ± 0.34 | 21.95 ± 0.58 |
| 10–12 | 26.09 ± 0.10 | 49.58 ± 0.19 | 24.32 ± 0.29 |
| 12–14 | 25.91 ± 0.03 | 49.23 ± 0.06 | 24.85 ± 0.09 |
| 14–16 | 25.96 ± 0.26 | 49.34 ± 0.49 | 24.69 ± 0.75 |
| 16–18 | 28.03 ± 0.25 | 47.65 ± 0.44 | 24.32 ± 0.69 |
| 18–20 | 25.72 ± 0.19 | 48.86 ± 0.36 | 25.40 ± 0.55 |
| 20–22 | 23.76 ± 0.21 | 49.89 ± 0.45 | 26.35 ± 0.66 |
| 22–24 | 25.42 ± 0.01 | 48.29 ± 0.02 | 26.29 ± 0.03 |
| 24–26 | 25.35 ± 0.19 | 48.17 ± 0.36 | 26.47 ± 0.55 |
| 26–28 | 23.13 ± 0.14 | 48.56 ± 0.28 | 28.31 ± 0.42 |
| 28–30 | 21.47 ± 0.23 | 49.37 ± 0.53 | 29.16 ± 0.78 |
| Stoichiometry | 22.67 | 45.33 | 32.00 |

The mole fractions of the PMN ($Pb(Mg_{1/3}Nb_{2/3})O_3$) and PT for different samples calculated from the ICP-MS data are shown in Figure 5. The PMN content decreases along the axial direction from the bottom to the top, while the PT content increases from 21 mol% to 29 mol%, which is consistent with that of the previous reports [22–24]. These results can be explained by the phase diagram of PMN-PT and the solidification law of the binary solid solution as follows: the PMN crystallizes firstly from the melt because of the higher freezing point of the PMN, which results in the higher content of the PMN at the bottom and the higher PT content in the liquid.

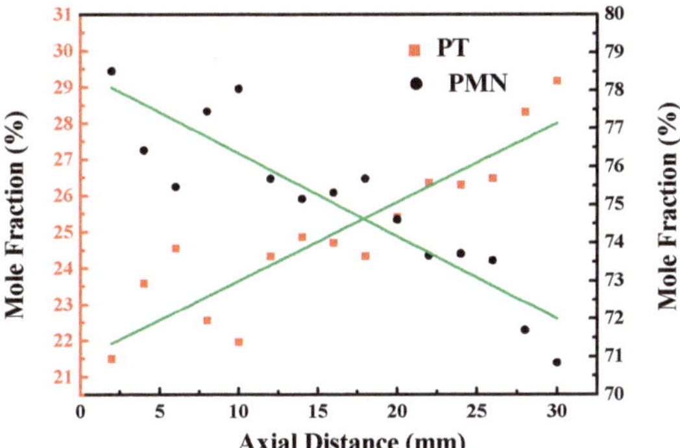

**Figure 5.** The distribution of the PMN and PT molar fractions along the axial direction: points represent the experimental data and solid lines represent the fitting.

The distribution of the molar ratio Nb/Mg along the axial direction is shown in Figure 6. It is seen that the molar ratio Nb/Mg exhibits an increasing trend from the bottom to the top. The molar ratio Nb/Mg fluctuates greatly at the initial stage of the crystal growth. During the crystal growth, it approaches to 2 in the middle part, and even >2 at the top boule. Theoretically, there are two $Nb^{5+}$ near a $Mg^{2+}$ in the lattice or melt to balance the valence state [31], namely $(Mg_{1/3}Nb_{2/3})^{4+}$. However, $Nb^{5+}$ and $Mg^{2+}$ in the actual lattice occupancy are not completely subject to theory due to the segregation.

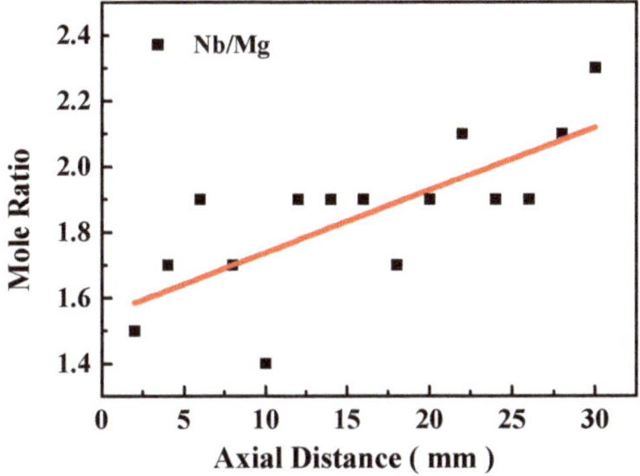

**Figure 6.** The molar ratio of Nb and Mg along the axial direction: points represent the experimental data and dashed lines represent the fitting.

The measured distance dependence of the electrical properties for the specimens along the axial direction is shown in Figure 7. From Figure 7a, we can see the permittivity firstly decreases at the bottom of the crystal and then tends to stabilize at the middle of the crystals. In addition, a sharp rise process for the dielectric constant is observed at the top. The distribution of dielectric loss exhibits a similar tendency, except for the sharp decrease at the top. The sharp variation of the dielectric properties at the top could originate from the presence of cracks. The piezoelectric constant $d_{33}$ increases gradually from 350 pC/N to 850 pC/N along the axial direction from the bottom to the top, as shown in Figure 7b. The coercive field $E_c$ and remnant polarization $P_r$ verse the measured distance for the PMN-32PT crystals are shown in Figure 7c,d. It is seen that the coercive field at room temperature fluctuates from 2 kV/cm to 3 kV/cm along the axial direction of the crystal. The remnant polarization of the crystal increases gradually except for some fluctuations from the bottom to the top. These results indicate that the segregation affects the electric properties dominantly in the PMN-PT single crystal grown by Bridgman method. In Figures 4 and 7a, it is obvious that the variation tendency of Ti content along the axial direction is opposite to that of the dielectric properties. Differently, the variations of the piezoelectric coefficient and remnant polarization are consistent with that of the PT content, as shown in Figures 5 and 7b,d. It is well known that it is multi-phase coexistence at the MPB composition for PMN-PT, in which the spontaneous polarization orientations increase. Therefore, the switch of domains and the motion of domain walls are easy under the external electric field, which makes it beneficial to obtain the ultrahigh piezoelectric constant and remnant polarization. The PT content increases gradually and approaches the MPB composition which is the reason for the increase of the piezoelectric coefficient and remnant polarization along the axial direction.

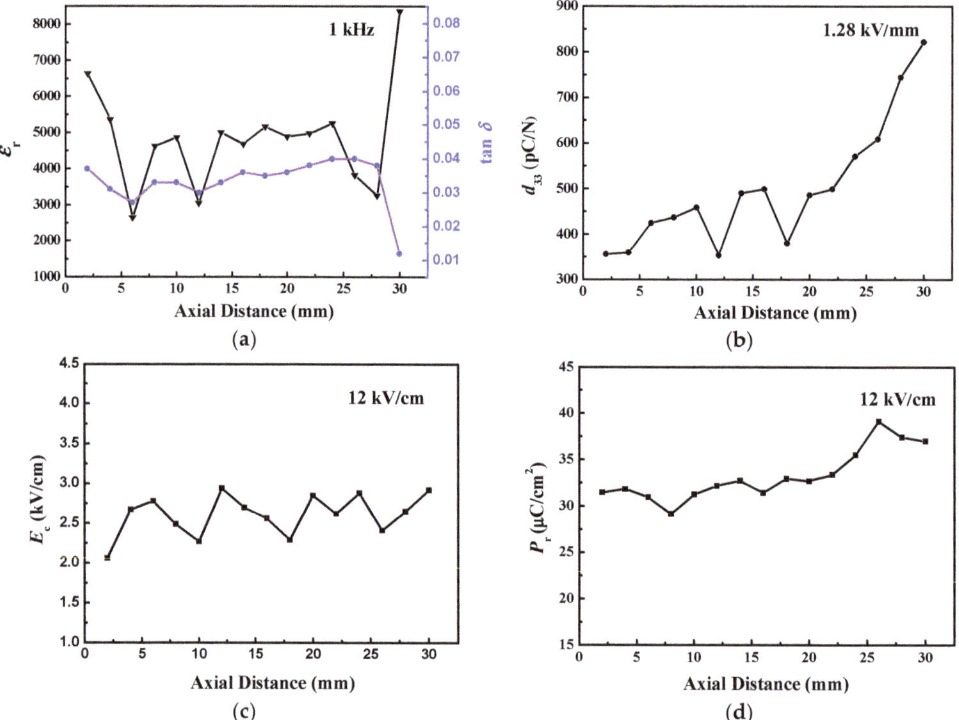


**Figure 7.** The variation of the electric properties along the axial direction: (**a**) the variation of the permittivity $\varepsilon$ and loss $\tan\delta$ at 1 kHz; (**b**) the variation of the piezoelectric constant $d_{33}$ poled under 1.28 kV/mm; (**c**) the variation of the coercive field Ec; (**d**) the variation of the remnant polarization Pr.

## 3.3. Radial Distribution

The mass fraction of different elements in a PMN-32PT single crystal along the radial direction is calculated and plotted in Figure 8. It is demonstrated that the Nb and Mg content increases by 0.71% 0.54%, respectively. By contrast, the content of Ti is relatively stable and increases only by 0.18%. The segregation of the radial direction is attributed to an uneven growth interface and the convection near growth interface [32–35], which is different from the segregation of the axial direction caused by the solute redistribution.

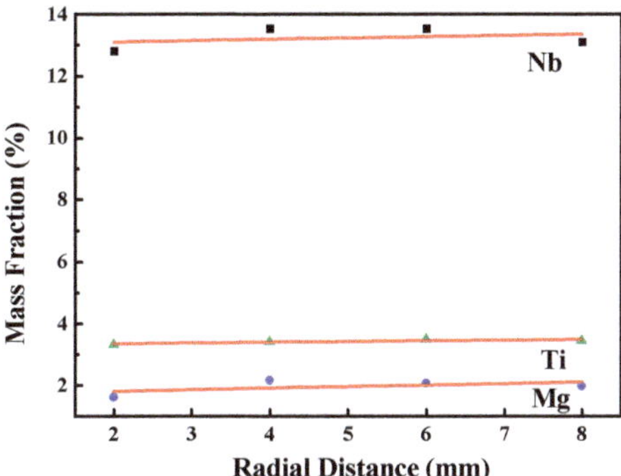

**Figure 8.** The composition distribution along the radial direction of PMN-32PT: points represent experimental data and the solid lines represent the fitting.

The variation of PMN and PT along the radial direction are presented in Figure 9. The mole fraction of PT nearly remains a constant of 24%, which means that PT is insensitive to the component segregation in the radial direction. Figure 10 illustrates the variation of Nb/Mg along the radial direction. The value decreases firstly and then increases slightly.

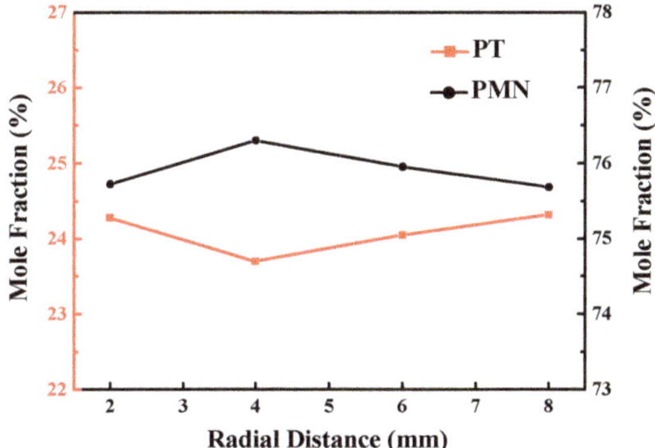

**Figure 9.** The distribution of PMN and the PT molar fraction along the radial direction.

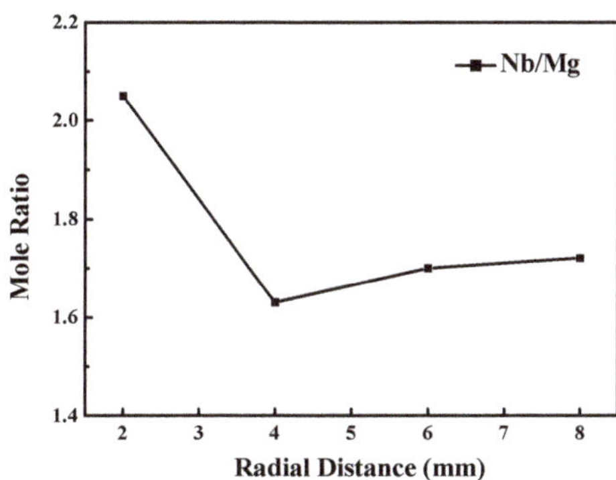

**Figure 10.** The distribution of the molar ratio of Nb and Mg along the radial direction.

The measured distance dependence of the electrical properties for the specimens along the radial direction is shown in Figure 11. The permittivity and piezoelectric constant decrease firstly and then increase from the sample X1 to X4, which are similar to that of Nb/Mg as shown in Figures 10 and 11a,b. The coercive field increases firstly and then decreases, which varies from the range of 2 kV/cm to 3 kV/cm shown in Figure 11c. The remnant polarization is almost maintained at 30–33 μC/cm². The relationship among the permittivity, piezoelectric constant and remnant polarization of the radial samples are basically obedient to the formula reported in the literature: $d_{33} = 2\varepsilon_0\varepsilon_r P_r Q_{11}$ [13,36]. The dependence of the dielectric constant, piezoelectric constant and coercive field on Nb/Mg can be explained as follows: as the Nb/Mg decreases, the concentration of the oxygen vacancies increases in the lattice, which pinches the domain and restrains the switch of the domain. As a result, the coercive field increases and the dielectric constant and piezoelectric constant decrease; and vice versa.

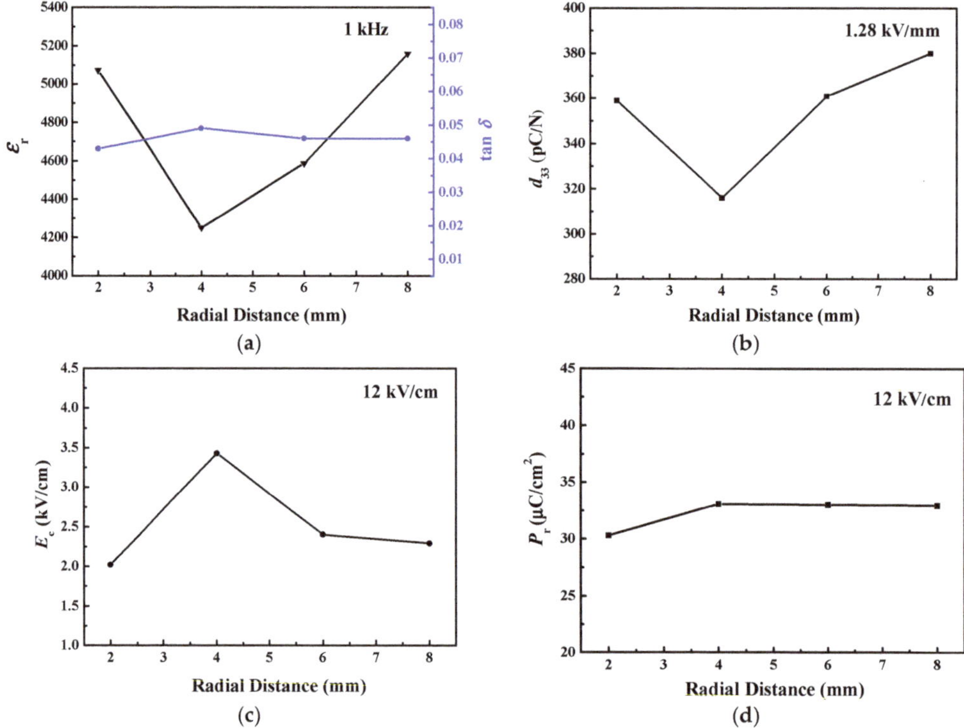

**Figure 11.** The variation of the electric properties along the radial direction: (**a**) the variation of permittivity $\varepsilon$ and loss tan$\delta$ at 1 kHz; (**b**) the variation of the piezoelectric constant $d_{33}$ poled under 1.28 kV/mm; (**c**) the variation of the coercive field $E$c; (**d**) the variation of the remnant polarization $P$r.

## 4. Conclusions

The single crystal with the nominal compositional PMN-32PT was grown using the Bridgman method. The distribution of the elements along the axial direction could be described by the effective segregation coefficient $k$. The effective segregation coefficients $k$ were 1.20, 1.03 and 0.82 for Mg, Nb and Ti, respectively. The relationship between the composition and electrical properties of the PMN-PT crystals was investigated. It was indicated that the piezoelectric constant and remnant polarization were consistent with the distribution of PT along the axial direction. The electrical properties were attributed to the distribution of the Nb and Mg along the radial direction.

**Author Contributions:** Writing-review & editing, Z.X.; writing-original draft, S.W.; formal analysis, P.F.; X.L.; investigation, W.L.; data curation, A.H.

**Funding:** This work was supported by the National Natural Science Foundation of China (Grant No. 51472197), the Key Laboratory of Optoelectronic Materials Chemistry Physics, Chinese Academy of Science (Grant No. 2016DP173016) and and Foundamental Research Foundation of XATU of china (Grant No. XAGDXJJ16020).

**Conflicts of Interest:** The authors declare no conflict of interest.

## References


1. Xue, A.X.; Fang, C.; Wang, C.; Jia, Y.M.; Liu, Y.S.; Luo, H.S. Uniaxial stress-induced ferroelectric depolarization in <001>-oriented 0.72Pb(Mg$_{1/3}$Nb$_{2/3}$)O$_3$-0.28PbTiO$_3$ single crystal. *J. Alloys Compd.* **2015**, *647*, 14–17. [CrossRef]
2. Li, Z.R.; Xu, Z.; Xi, Z.Z.; Cao, L.H.; Yao, X. Dielectric loss anomalies of 0.68PMN-0.32PT single crystal and ceramics at cryogenic temperature. *J. Electroceram.* **2008**, *21*, 279–282. [CrossRef]
3. Wang, Z.; Zhang, R.; Sun, E.W.; Cao, W.W. Temperature dependence of electric-field-induced domain switching in 0.7Pb(Mg$_{1/3}$Nb$_{2/3}$)O$_3$-0.3PbTiO$_3$ single crystal. *J. Alloys Compd.* **2012**, *527*, 101–105. [CrossRef] [PubMed]
4. Sun, E.W.; Cao, W.W. Relaxor-based ferroelectric single crystals: Growth, domain engineering, characterization and applications. *Prog. Mater. Sci.* **2014**, *65*, 124–210. [CrossRef] [PubMed]
5. Li, F.; Zhang, S.; Xu, Z.; Wei, X.Y.; Luo, J.; Shrout, T.R. Composition and phase dependence of the intrinsic and extrinsic piezoelectric activity of domain engineered $(1-x)$Pb(Mg$_{1/3}$Nb$_{2/3}$)O$_{3-}$$x$PbTiO$_3$ crystals. *J. Appl. Phys.* **2010**, *108*, 074106. [CrossRef] [PubMed]
6. Guerra, J.D.L.S.; Lente, M.H.; Eiras, J.A. Non-linear dielectric properties in based-PMN relaxor ferroelectrics. *J. Eur. Ceram. Soc.* **2017**, *27*, 4033–4036. [CrossRef]
7. Zhou, Q.F.; Lam, K.H.; Zheng, H.R.; Qiu, W.B.; Shung, K.K. Piezoelectric single crystal ultrasonic transducers for biomedical applications. *Prog. Mater. Sci.* **2014**, *66*, 87–111. [CrossRef]
8. Shkuratov, S.I.; Baird, J.; Antipov, V.G.; Talantsev, E.F.; Chase, J.B.; Hackenberger, W.; Luo, J.; Jo, H.R.; Lynch, C.S. Ultrahigh energy density harvested from domain-engineered relaxor ferroelectric single crystals under high strain rate loading. *Sci. Rep.* **2017**, *7*, 46758. [CrossRef]
9. Guo, M.S.; Dong, S.X.; Ren, B.; Luo, H.S. A double-mode piezoelectric single-crystal ultrasonic micro-actuator. *IEEE Trans. Ultrason. Ferroelectr. Freq. Control* **2010**, *57*, 2596–2600.
10. Xu, G.S.; Luo, H.S.; Wang, P.C.; Xu, H.Q.; Yin, Z.W. Ferroelectric and piezoelectric properties of novel relaxor ferroelectric single crystals PMNT. *Chin. Sci. Bull.* **2000**, *45*, 491–495. [CrossRef]
11. Tian, J.; Han, P.D. Crystal growth and property characterization for PIN-PMN-PT ternary piezoelectric crystals. *J. Adv. Dielectr.* **2014**, *4*, 1350027. [CrossRef]
12. Tian, J.; Han, P.D.; Huang, X.L.; Pan, H.X.; Carroll, J.F.; Payne, D.A. Improved stability for piezoelectric crystals grown in the lead indium niobate-lead magnesium niobate-lead titanate system. *Appl. Phys. Lett.* **2007**, *91*, 222903. [CrossRef]
13. Li, F.; Lin, D.B.; Chen, Z.B.; Cheng, Z.X.; Wang, J.L.; Li, C.C.; Xu, Z.; Huang, Q.W.; Liao, X.Z.; Chen, L.Q.; et al. Ultrahigh piezoelectricity in ferroelectric ceramics by design. *Nature. Mater.* **2018**, *17*, 349–354. [CrossRef] [PubMed]
14. Xu, J.L.; Deng, H.; Zeng, Z.; Zhang, Z.; Zhao, K.Y.; Chen, J.W.; Nakamori, N.M.; Wang, F.F.; Ma, J.P.; Li, X.B.; et al. Piezoelectric performance enhancement of Pb(Mg$_{1/3}$Nb$_{2/3}$)O$_3$-0.25PbTiO$_3$ crystals by alternating current polarization for ultrasonic transducer. *Appl. Phys. Lett.* **2018**, *112*, 182901. [CrossRef]
15. Wang, D.; Yuan, G.L.; Luo, H.S.; Li, J.F.; Viehland, D.; Wang, Y.J. Structural origin of room temperature poling enhanced piezoelectricity in modified Pb(Mg$_{1/3}$Nb$_{2/3}$)O$_3$-30%PbTiO$_3$ crystals. *J. Am. Ceram. Soc.* **2017**, *100*, 4938–4944. [CrossRef]
16. Jiao, S.; Tang, Y.X.; Zhao, X.Y.; Wang, T.; Duan, Z.H.; Wang, F.F.; Sun, D.Z.; Luo, H.S.; Shi, W.Z. Growth and electrical properties of epitaxial 0.7Pb(Mg$_{1/3}$Nb$_{2/3}$)O$_3$-0.3PbTiO$_3$ thin film by pulsed laser deposition. *J. Mater. Sci.* **2018**, *29*, 6779–6784. [CrossRef]
17. Li, F.; Zhang, S.J.; Yang, T.N.; Xu, Z.; Zhang, N.; Liu, G.; Wang, J.L.; Wang, J.L.; Cheng, Z.X.; Ye, Z.G.; et al. The origin of ultrahigh piezoelectricity in relaxor-ferroelectric solid solution crystals. *Nat. Common.* **2016**, *7*, 13807. [CrossRef]

18. Hu, W.H.; Xi, Z.Z.; Fang, P.Y.; Long, W.; Li, X.J.; Bu, Q.Q. A novel poling technique to obtain excellent piezoelectric properties of $Pb(Mg_{1/3}Nb_{2/3})O_3$-$0.32PbTiO_3$ single crystals. *J. Mater. Sci. Mater. Electron.* **2015**, *26*, 3282–3286. [CrossRef]

19. He, A.G.; Xi, Z.Z.; Li, X.J.; Long, W.; Fang, P.Y.; Zhao, J.; Yu, H.N.; Kong, Y.L. Optical properties of $Ho^{3+}$- and $Ho^{3+}/Yb^{3+}$-modified PSN-PMN-PT crystals. *Mater. Lett.* **2018**, *219*, 64–67. [CrossRef]

20. Long, W.; Chu, X.; Xi, Z.Z.; Fang, P.Y.; Li, X.J.; Cao, W.W. Growth and property enhancement of $Er^{3+}$-doped $0.68Pb(Mg_{1/3}Nb_{2/3})O_3$-$0.32PbTiO_3$ single crystal. *J. Rare. Earth.* **2018**, *36*, 832–837. [CrossRef]

21. Xi, Z.Z.; He, A.G.; Fang, P.Y.; Li, X.J.; Long, W. Electric and optical properties of $Er^{3+}$- and $Er^{3+}/Yb^{3+}$-modified PSN-PMN-PT crystals. *J. Alloys Compd.* **2017**, *722*, 375–380. [CrossRef]

22. Luo, H.S.; Xu, G.S.; Xu, H.Q.; Wang, P.C.; Yin, Z.W. Compositional homogeneity and electrical preperties of lead magnesium niobate titanate single crystals grown by a Modified Bridgman technique. *J. Appl. Phys.* **2000**, *39*, 5581–5585. [CrossRef]

23. Benayad, A.; Sebald, G.; Lebrun, L.; Guiffard, B.; Pruvost, S.; Guyomar, D.; Beylat, L. Segregation study and segregation modeling of Ti in $Pb[(Mg_{1/3}Nb_{2/3})_{0.60}Ti_{0.40}]O_3$ single crystal grown by Bridgman method. *Mater. Res. Bull.* **2006**, *41*, 1069–1076. [CrossRef]

24. Zawilski, K.T.; Custodio, M.C.C.; Demattei, R.C.; Lee, S.G.; Monteiro, R.G.; Odagawa, H.; Feigelson, R.S. Segregation during the vertical Bridgman growth of lead magnesium niobate-lead titanate single crystals. *J. Cryst. Growth* **2003**, *258*, 353–367. [CrossRef]

25. Guo, Z.Q.; Fu, T.; Fu, H.Z. Crystal Orientation Measured by XRD and Annotation of the Butterfly Diagram. *Mater. Charact.* **2000**, *44*, 431–434. [CrossRef]

26. Guo, Z.Q.; Jin, L.; Li, F.; Bai, Y. Applications of the rotating orientation XRD method to oriented materials. *J. Appl. Phys. D* **2009**, *42*, 012001. [CrossRef]

27. Guo, Z.Q.; Fu, T.; Wang, N.; Fu, H.Z. A sample XRD method for determining crystal orientation and its distribution. *J. Inorg. Mater.* **2002**, *17*, 460–464.

28. Guo, Z.Q.; Li, F.; Xu, Z. Application of new equipment to determine the orientation of single crystal by XRD. *Lab. Sci.* **2011**, *14*, 92–96.

29. Zhang, Y.Y.; Li, X.B.; Liu, D.A.; Zhang, Q.H.; Wang, W.; Ren, B.; Lin, D.; Zhao, X.Y.; Luo, H.S. The compositional segregation, phase structure and properties of $Pb(In_{1/2}Nb_{1/2})O_3$-$Pb(Mg_{1/3}Nb_{2/3})O_3$-$PbTiO_3$ single crystal. *J. Cryst. Growth* **2011**, *318*, 890–894. [CrossRef]

30. Zhang, S.J.; Li, F. High performance ferroelectric relaxor-$PbTiO_3$ single crystals: Status and perspective. *J. Appl. Phys.* **2012**, *111*, 031301. [CrossRef]

31. Song, K.X.; Li, Z.R.; Guo, H.S.; Xu, Z.; Fan, S.J. Compositional segregation and electrical properties characterization of [001]-and [011]-oriented co-growth $Pb(In_{1/2}Nb_{1/2})O_3$-$Pb(Mg_{1/3}Nb_{2/3})O_3$-$PbTiO_3$ single crystal. *J. Appl. Phys.* **2018**, *123*, 154107. [CrossRef]

32. Ganaoui, M.E.; Bontoux, P. Gravity effects on solidification: The case of an unsteady melt affecting the growth interface. *Adv. Space Res.* **1999**, *24*, 1423–1426. [CrossRef]

33. Volkov, P.K.; Zakharov, B.G.; Serebryakov, Y.A. Numerical and experimental investigations of convection and heat/mass transfer effect in melts on inhomogeneity formation during Ge crystal growth by the Bridgman method. *J. Cryst. Growth* **1999**, *204*, 475–486. [CrossRef]

34. Yang, C.; Xu, Q.Y.; Liu, B.C. Study of dendrite growth with natural convection in superalloy directional solidification via a multiphase-field-lattice Boltzmann model. *Comp. Mater. Sci.* **2019**, *158*, 130–139. [CrossRef]

35. Antar, B.N. Convective instabilities in the melt for solidifying mercury cadmium telluride. *J. Cryst. Growth* **1991**, *113*, 92–102. [CrossRef]

36. Lines, M.E.; Glass, A.M. *Principles and Applications of Ferroelectrics and Related Materials*; Oxford Univ. Press: Oxford, UK, 1977.

*Review*

# Prospective of (BaCa)(ZrTi)O$_3$ Lead-free Piezoelectric Ceramics

**Wenfeng Liu *, Lu Cheng and Shengtao Li**

State Key Laboratory of Electrical Insulation and Power Equipment, Xi'an Jiaotong University, Xi'an 710049, China; lu.cheng@stu.xjtu.edu.cn (L.C.); sli@xjtu.edu.cn (S.L.)
* Correspondence: liuwenfeng@xjtu.edu.cn

Received: 11 February 2019; Accepted: 13 March 2019; Published: 26 March 2019

**Abstract:** Piezoelectric ceramics is a functional material that can convert mechanical energy into electrical energy and vice versa. It can find wide applications ranging from our daily life to high-end techniques and dominates a billion-dollar market. For half a century, the working horse of the field has been the polycrystalline PbZr$_{1-x}$Ti$_x$O$_3$ (PZT), which is now globally resisted for containing the toxic element lead. In 2009, our group discovered a non-Pb piezoelectric material, (BaCa)(ZrTi)O$_3$ ceramics (BZT-BCT), which exhibits an ultrahigh piezoelectric coefficient $d_{33}$ of 560–620 pC/N. This result brought extensive interest in the research field and important consequences for the piezoelectric industry that has relied on PZT. In the present paper, we review the recent progress, both experimental and theoretical, in the BZT-BCT ceramics.

**Keywords:** piezoelectric; ceramic; lead-free

## 1. Introduction

Piezoelectricity refers to the phenomenon of interconversion between mechanical energy and electrical energy, which yields a mechanical-stress-induced polarization or an electrical-field-induced strain. Such ability of energy conversion enables piezoelectric materials to be widely used in devices such as sensors, actuators, transducers, etc. [1,2]. For more than half a century, Pb-based piezoelectric ceramics (e.g. PbZr$_{1-x}$Ti$_x$O$_3$, PbMg$_x$Nb$_{1-x}$-PbTiO$_3$ and PbZn$_x$Nb$_{1-x}$-PbTiO$_3$) have dominated the area of most applications. However, the use of Pb-based materials is restricted by increasingly tight regulations due to its high toxicity [3,4]. This arouses extensive investigations on the mechanism of high piezoelectricity in Pb-based materials and the exploration of Pb substitutes [5–8].

In 2009, a large piezoelectric performance with $d_{33}$ of 620 pC/N was observed for 0.5Ba(Zr$_{0.2}$Ti$_{0.8}$)O$_3$-0.5(Ba$_{0.7}$Ca$_{0.3}$)TiO$_3$ composites (BZT-50BCT) [9]. Later, the modified BZT-BCT composite ceramics showed a higher $T_c \sim 114\,°C$ than BZT-0.53BCT [10]. By optimizing the poling conditions, BZT-BCT ceramics combines a large piezoelectric performance with $d_{33}$ of 630 pC/N and the planar electromechanical factor of 56% [11]. Yang et al. prepared the lead-free BZT-BCT ceramics by sol-gel technique [12], whose maximum permittivity was above 9000 with a maximum converse piezoelectric coefficient ($d_{33}$*) of 400 pm/V.

In this review, we summarize the recent progresses on BZT-BCT piezoelectric ceramics by different doping mechanisms that may offer some thoughts on the future improvement of BZT-BCT and even other piezoelectric ceramics. Further, the potential application of BZT-BCT ceramics are presented, including electrocaloric effect, fluorescence and energy storage. Based on the current achievements, we also propose some prospects, which may provide new directions on the development of BZT-BCT ceramics.

## 2. MPB Strategy

In both Pb-based and Pb-free piezoelectric systems, the common solution to promote piezoelectric performance is to place materials at their phase transition boundaries, either a paraelectric to ferroelectric phase boundary or multi ferroelectric phases coexisting boundary (including the most famous morphotropic phase boundary, i.e. MPB), since the instability of the polarization at phase boundaries allows a significant polarization variation under an external stress or electric field. Despite the intense interests in the phase coexisting strategy, the key to understanding the consequent high piezoelectricity in BZT-$x$BCT ceramics is still in dispute. For instance, we initially proposed the coexistence of tetragonal (T) and rhombohedral (R) symmetry at MPB, evidenced by the synchrotron X-ray diffraction (XRD) results from Ehmke [13] and transmission electron microscope (TEM) results from Gao [14]; however, soon after, the discovery of an intermediate orthorhombic (O) phase was found based on synchrotron XRD results from Keeble [15,16] and temperature spectrum of dielectric permittivity from Damjanovic [17].

In addition, the phenomenological Landau–Devonshire model suggests that the reduction of polarization anisotropy is responsible for enhanced piezoelectric response approaching to MPB [18,19]. Acosta calculated the anisotropy energy of a sixth-order Landau potential formulated for the BZT-$x$BCT system and found that the anisotropy energy approaches zero near the O-R rather than the T-O phase boundary. They thus attributed the best piezoelectric property found at the T-O phase boundary to two other factors, i.e., higher degree of poling and increased elastic softening [20]. Ke used the energy barrier along the minimum energy pathway on the free energy surface for direct domain switching to quantitatively measure the degree of polarization anisotropy and suggested that the polarization anisotropy at the T-O phase boundary was the smallest [21].

In most MPB systems (e.g., PMN-PT and PZN-PT), MPB is temperature dependent, similar to the BZT-BCT system, but this does not render such systems useless. Many important applications of this system have been found. The widely used solution to achieve temperature stability is to choose a composition slightly away from the MPB, so that the properties are no longer sensitive over the ambient temperature range. However, this is at the expenses of a slight reduction of piezoelectric properties. It is the same situation in the BZT-BCT ceramics. As shown in Figure 1, by choosing BZT-45BCT, a composition slightly off MPB composition (50BCT), $d_{33}$ becomes almost temperature independent in the room temperature range (20–40 °C). This temperature stability is achieved with some sacrifices of $d_{33}$, but the $d_{33}$ (~360 pC/N) is still much higher than many other non-Pb piezoelectric materials.

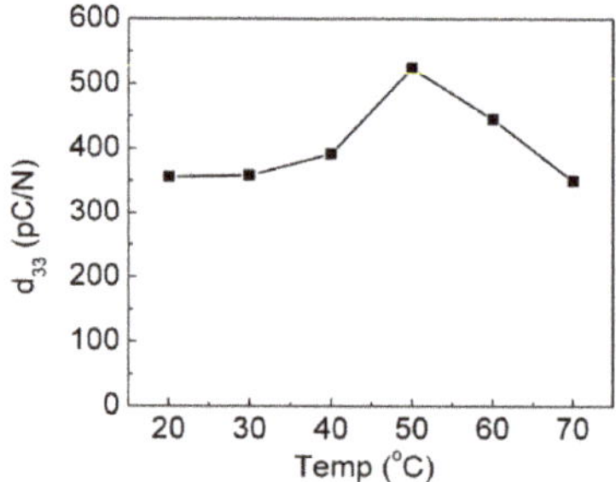

**Figure 1.** Temperature dependence of $d_{33}$ of BZT-45BCT.

## 3. Approach to Tailor the Piezoelectric Performance of BZT-BCT

The chemical modification is the most efficient and widely accepted way to tailor the piezoelectric performance of piezoceramics. Table 1 summarizes the most attractive achievements of both pure BZT-BCT ceramic and doped ceramics. Based on the mechanism of chemical modification, the composition optimization can be categorized into improving microstructure by sintering aids and substitution doping, and the most improved piezoelectric performances of each doping element can be found in Figure 2.

**Table 1.** Piezoelectric performance of the BZT-BCT based ceramics.

| Compositions | $d_{33}$ (pC/N) | $k_p$ | $\varepsilon_r$ | $T_c$ (°C) | Reference |
|---|---|---|---|---|---|
| BZT-50BCT | 650 | 53% | 4500 | 85 | [22] |
| | 630 | 56% | | 90 | [11] |
| | 620 | | 3060 | 93 | [23] |
| | 572 | 57% | 4821 | 94.8 | [24] |
| | 546 | 65% | 4050 | | [25] |
| | 464 | | 2938 | | [26] |
| BZT-50BCT-0.08 wt% ZnO | 603 | | | | [27] |
| BZT-50BCT-0.04 wt% CeO$_2$ | 600 | 51% | 4843 | | [28] |
| BZT-50BCT-0.1 wt% CeO$_2$ | 565 | 52% | 3860 | | [29] |
| BZT-50BCT-0.2 wt% Sr(Cu$_{1/3}$Ta$_{2/3}$)O$_3$ | 577 | ####### | | 97 | [30] |
| 99.2 mol% (BZT-50BCT)-0.8 mol% BiAlO$_3$ | 568 | 54% | 3375 | 72 | [31] |
| BZT-50BCT-0.06 wt% Y$_2$O$_3$ | 560 | 53% | | 95 | [32] |
| BZT-50BCT-0.1% Sb$_2$O$_3$ | 556 | 52% | 3985 | | [33] |
| (Ba$_{0.82}$Sr$_{0.03}$Ca$_{0.15}$)(Zr$_{0.1}$Ti$_{0.9}$)O$_3$ | 534 | ####### | | 84 | [34] |
| BZT-50BCT-0.06 mol% ZnO | 521 | ####### | | | [35] |
| BZT-50BCT-1 mol% Sn-1 mol% Sr | 514 | ####### | | | [36] |
| BZT-50BCT-0.3 wt% Li$_2$CO$_3$ | 512 | 49% | 4394 | 79.6 | [37] |
| BZT-50BCT-0.04 wt% CuO | 510 | 45% | 3762 | 95 | [38] |
| BZT-50BCT-0.5 mol% SiO$_2$ | 500 | | | | [39] |

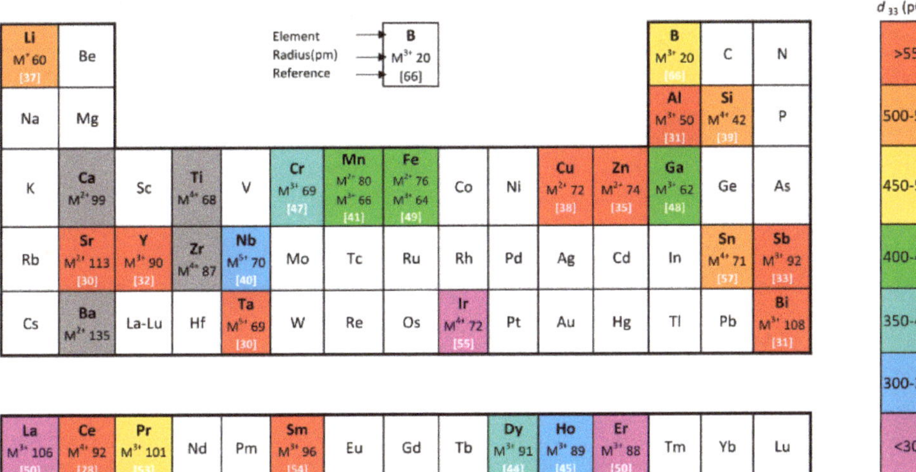


**Figure 2.** Piezoelectric coefficient ($d_{33}$) of the point-defects modified BZT-BCT ceramics.

### 3.1. Substitution Doping

Ferroelectrics are rarely used in a chemically pure form and doping is always employed with the goal of tailoring the properties for specific applications. Different doping usually brings about different consequence, such as acceptor doping to obtain low dielectric losses and donor doping to achieve high piezoelectric coefficients. For the equivalent doping, it is usually employed to modify the Curie temperature and introduce the compositional disorder, which may also greatly affect the piezoelectric performance. As shown in Figure 2, the radius of $Ba^{2+}$ and $Ca^{2+}$ are 135 pm and 99 pm, respectively, while $Ti^{4+}$ and $Zr^{4+}$ are 68 pm and 87 pm, respectively. Usually, large ions will occupy the A-site, and small ions will occupy the B-site, and intermediate ions will occupy both sites with different ratios of the perovskite structure.

### 3.1.1. Acceptor Doping

An acceptor dopant has a lower oxidation number than that of the host cation. This will create oxygen vacancies owing to the charge compensation, and the oxygen vacancies can contribute to the mass transport and improve the density during sintering process [27,35,40].

$Zn^{2+}$ has a radius of 74 pm and prefers to replace smaller ions in B-site. JG Wu pointed out that the $Zn^{2+}$ substitution into the $(Ti, Zr)^{4+}$ site results in both structure disorder and lattice distortion. The presence of oxygen vacancies helps the mass transport during sintering, which is responsible for the enhanced grain growth as the ZnO content increases. Besides, the tricritical point of these ceramics was shifted to room temperature while the Curie temperature decreased simultaneously by the introduction of ZnO. Macroscopically, the BZT-BCT ceramics have $\varepsilon_r \sim 4500$ and $\tan\delta < 1.5\%$. BZT-BCT ceramic with 0.06 mol.% ZnO demonstrates an enhanced electrical behavior with $d_{33} \sim 521$ pC/N, $k_p \sim 47.8\%$, and $2P_r \sim 19.37$ $\mu C/cm^2$, owing to the room-temperature tricritical point induced by doping with ZnO [35]. Similar results were also reported by Zhao [27]. In Zhao's work, the BZT-BCT ceramics with 0.08 wt% ZnO show maximum remnant polarization and spontaneous polarization ($P_r = 10.14$ $\mu C/cm^2$, $P_s = 19.68$ $\mu C/cm^2$). At the same time, the giant piezoelectric coefficient of $d_{33}$ = 603 pC/N and high planar electromechanical coupling factor of $k_p = 0.56$ were also obtained for the samples. Besides, they revealed that further raising ZnO content would cause partial $Zn^{2+}$ ions occupying A-site, resulting in decrease in lattice parameters. Consequently, excessive $Zn^{2+}$ are effective in reducing the poling state and in depressing the domain switching of BCZT–$x$Zn ceramics; thus, they degrade the dielectric, ferroelectric and piezoelectric properties.

Mn as an interesting impurity exhibits multivalence states ($Mn^{2+}$, $Mn^{3+}$, and $Mn^{4+}$), and has been extensively used to substitute the A- and/or B-site ions of ferroelectrics for tailoring electrical properties [41]. Typically, it manifests itself as $Mn^{2+}$ when sintering at high temperature above 850 °C. Meng et al. reported that the addition of 0.25 mol% $MnO_2$ promotes grain growth, improves the ferroelectricity of the ceramics and strengthens ferroelectric tetragonal–ferroelectric orthorhombic phase transition near 40 °C [42]. Such ceramics exhibit the optimum piezoelectric properties ($d_{33} = 306$ pC/N and $k_p = 42.2\%$, respectively). Excess $MnO_2$ inhibits the grain growth and degrades the ferroelectric and piezoelectric properties of the ceramics. Wu et al. reported similar results: the 0.15 wt% MnO doped BZT-BCT ceramics have optimum electrical properties: $d_{33} = 382$ pC/N, $k_p = 44.5\%$, $\varepsilon_r = 2611$, and $\tan\delta = 0.63\%$. Besides, they also observed the double hysteresis loop when MnO dopant exceeds 0.3 wt%, which indicates the reversible domain switching dominating by the acceptor defect dipoles [43].

### 3.1.2. Donor Doping

A donor dopant has a higher oxidation number than that of the host cation. This will create cationic vacancies, either A-site or B-site vacancies in the perovskite structure. The donor dopant may effectively reduce the Curie temperature and facilitate the domain switching behavior, characterized as slim hysteresis loops with small coercive field and hysteresis loss.

Li et al. reported that, with the introduction of the donor Dy at A-site, the BZT-BCT ceramics possessed improved temperature stability. Besides the high piezoelectric coefficient of $d_{33} = 366$ pC/N and planar electromechanical coupling factor of $k_p = 43.0\%$, the Dy-doped BZT-BCT ceramics exhibited stable electromechanical coupling coefficients over a common usage temperature range of 20–100 °C [44]. They also revealed similar results in Ho-doped BZT-BCT ceramics [45].

As another common donor dopant to A-site, $La^{3+}$ is often employed to enhance the electrical or mechanical responds under small AC field while reducing the loss in ferroelectric ceramics. Sun et al. reported that a small amount of $La^{3+}$ (~0.15%) resulted in an increase of $d_{33}$ at 50 °C due to the coexistence of orthorhombic and tetragonal phases. Besides, the 0.15 mol% La doped BZT-BCT ceramics turned to the semi-conductor and showed positive temperature coefficient (PTC) behavior [46]. Other donor dopants such as $Cr^{3+}$ and $Ga^{3+}$ also showed excellent improvements of $d_{33}$[47,48].

Besides the mono-element doped ceramics, compound doped BZT-BCT are also investigated. Wu et al. investigated the $BiFeO_3$ doped BZT-BCT ceramics [49]. Here, $Bi^{3+}$ acted as the donor dopant at A-site while $Fe^{3+}$ as the acceptor at B-site. By addition of $BiFeO_3$, the grain size becomes smaller, and these ceramics become denser. The 0.2 mol% $BiFeO_3$ doped BZT-BCT ceramics demonstrated an improved piezoelectric behavior ($d_{33}$ ~ 405 pC/N and $k_p$~0.44). Tian et al. added $Er^{3+}$ and $La^{3+}$ into the BZT-BCT ceramics and found $Er^{3+}$ first substituted A sites and then B sites in the matrix of $ABO_3$ structure. The elevated piezoelectric constant ($d_{33}$ ~ 200 pC/N) and receded mechanical quality factor ($Q_m$ ~ 70) at with 0.12% $La^{3+}$ doping and 0.2% $Er^{3+}$ doping showed "softening effect" by donor doping [50].

Extensive studies with other donor dopants were also carried on the BZT-BCT ceramics [30,51–54], as shown in Figure 2.

### 3.1.3. Equivalent Doping

Researchers have tried other doping with equivalent valence as the A-site or B-site to achieve MPB in the $BaTiO_3$ ceramics [55–60]. Zhou et al. designed a new Pb-free piezoelectric system $Ba(Hf_{0.2}Ti_{0.8})O_3$-$(Ba_{0.7}Ca_{0.3})TiO_3$, which yields high piezoelectricity with $d_{33}$ ~ 550 pC/N, comparable to that of the best Pb-based material PZT-5H ($d_{33}$ ~ 590 pC/N). Besides, their study suggests the non-isotropicity of polarization at triple point by precise detection of transitional thermal hysteresis. Previously, the isotropy at triple point was always taken as the basic assumption in related modeling work and understanding [56]. Xue et al. designed a similar Pb-free pseudo-binary system, $Ba(Sn_{0.12}Ti_{0.88})O_3$-$x(Ba_{0.7}Ca_{0.3})O_3$, characterized by a phase boundary starting from a critical triple point of a paraelectric cubic phase, ferroelectric rhombohedral, and tetragonal phases [57]. The optimal composition BTS-30BCT exhibits a high piezoelectric coefficient $d_{33}$ = 530 pC/N at room temperature.

For the equivalent doping of A-site, one of the most typical substitution is $Sr^{2+}$, which can contribute to the microstructure by increasing grainsize and density, and at the same time decrease $T_c$ [34,36,58,59]. As a result of fine microstructure and lower $T_c$, excellent piezoelectric properties with large $d_{33}$ of 534 pC/N was exhibited in Bai and Li's work when 0.03 mol $Sr^{2+}$ was doped [34]. $Sn^{4+}$ is another commonly used element locating at B-site to enhance the piezoelectric properties of lead-free piezoelectric ceramics, especially in BT based ceramics [57,60]. In Ding and Liu's work, the co-doping of $Sn^{4+}$ and $Sr^{2+}$ could lead to good piezoelectrical properties of $d_{33}$ = 514 pC/N and $k_p$ = 52.62%, although smaller grain size was obtained comparing with pure BZT-BCT ceramics, which is different from many other dopants that usually enhance the grain growth with low content [36].

### 3.2. Grain Size Effect and Sintering Aids

It is well known that grain size can have a significant influence on the properties of ferroelectric ceramics. For BZT-BCT based ceramics, the benefits to $d_{33}$ from large grain size are extremely obvious, as shown in Figure 2. Elements acting as sintering aids such as Ce, Cu, Si, Li, etc. present brilliant performance. As the grain size decreased to the micron level, the permittivity at room temperature increased. The increase in permittivity can be understood in terms of the twinning behavior of polycrystals with decreasing grain size [61].

Hao et al. fabricated various BZT-50BCT ceramics with different grain size ranging from 0.4 to 32 μm. As grain size decreases, the diffuse phase transition behavior is enhanced. As the grains grow up to more than 10 μm, samples exhibit good piezoelectric properties with $k_p$ > 0.48,

$k_t > 0.46$, $d_{33} > 470$ pC/N, and $d_{33}^* > 950$ pm/V. Besides, the increasing grain size effectively enhances the resistance to thermal depolarization [62].

In application, sintering aids are usually employed to enlarge the grain size and modify the microstructure to lower the sintering temperature and at the same time enhance the performance [28,29,37,63,64]. $SiO_2$, $Li^+$, and CuO are widely used as the sintering aid of ceramics and exhibited good contributions in BZT-BCT system. In Chen's work, the dopant of $Li^+$ reduced the sintering temperature of BZT-50BCT from 1540°C to 1400°C and piezoelectric performance with $d_{33} = 512$ pC/N and $k_p = 49\%$ were obtained [37]. In the work of co-doping of $CeO_2$ and $Li_2CO_3$, adding of $Li_2CO_3$ could largely decrease the sintering temperature from 1450°C to 1050°C comparing with the composition of $CeO_2$ dopant only [29]. Liu systematically studied the effect of $SiO_2$ dopant on the dielectric, ferroelectric and piezoelectric properties of the BZT-50BCT ceramics sintered at different temperatures and the results show that doping of $SiO_2$ could enhance $P_r$ and $d_{33}$ while reducing $E_c$. They obtained $d_{33}$ of 500 pC/N [39,65]. CuO was also found to be effective on decreasing sintering temperature and modifying the microstructure of BZT-BCT ceramics, thua enhanced piezoelectric properties were achieved due to the large grain size cause by CuO [30,46,54]. The co-doping of CuO and $B_2O_3$ can largely benefit the morphology and enhance the piezoelectric properties [66]. Besides, some other substitution doping elements can also contribute to the sintering process at the same time such as Zn and Mn. In Wu's work, the grain size of ZnO doped BZT-BCT ceramic increased and surface morphologies became denser with the help of ZnO [35]. Similarly, MnO could also help sintering process greatly and benefit the piezoelectric performance [43,67].

## 4. For other Applications

It is well accepted that the easy polarization extension and rotation mechanism (low energy barriers for polarization variation) are the two most important intrinsic factors contributing to the enhanced physical properties [68,69]. Besides the good piezoelectric performance, BZT-BCT also exhibits other good properties due to its easy polarization rotation and easy domain wall motion.

### 4.1. Electrocaloric Effect

Electrocaloric effect (ECE) is a phenomenon that the change in adiabatic temperature and/or entropy of a dielectric material is induced by the application and removal of an electric field due to the change in the dipolar state of the material. In general, the ECE is parameterized by the adiabatic temperature change ($\Delta T$), and the electrocaloric efficiency ($\Delta T/\Delta E$). The main technical challenge in lead-free ferroelectric bulk materials is to generate a giant electrocaloric temperature change $\Delta T$ under a relatively low electric field [70].

BZT-BCT has attracted much attention due to its relatively low coercive field and large polarization. Besides, the low Curie temperature also benefits the applications at room temperature. Sanlialp et al. reported that 0.65BZT-0.35BCT ceramics exhibited a $\Delta T$ of 0.33 °C and $\Delta T/\Delta E$ of 0.165 K mm kV$^{-1}$ at 65 °C under an electric field change $\Delta E = 20$ kV/cm [71]. G. Singha et al. reported large ECE in BZT-BCT ceramics [72]. They revealed that BZT-0.8BCT possessed an electro-caloric coefficient as high as 0.253 K mm/kV near tetragonal-to-cubic phase transition. They ascribed the high ECE to the higher polarization flexibility. Wang et al. performed direct measurements for BZT-BCT ceramics in non-adiabatic and non-equilibrium conditions. They found that BZT-0.7BCT showed the maximal ECE temperature change ($\Delta TECE$) of 0.55 K was recorded for an applied field of 40 kV/cm at T = 85 °C [73]. Besides the good performance, they also revealed a very large discrepancy between the indirectly estimated and the directly measured $\Delta T_{ECE}$. This was attributed to non-adiabatic conditions of the experiment resulting in a heat exchange with the environment.

### 4.2. Fluorescence

In recent years, ferroelectrics doped with rare earth elements have attracted great attention owing to their excellent multifunctional properties [74,75]. Multi-property coupling among the presence of an electric field, mechanical stress and photons can introduce many applications as piezoelectric ceramics, mechano-luminescence and electro-luminescence materials. Peng Du et al. investigated the fluorescence intensity ratio of green up conversion emissions at 525 and 550 nm in the temperature range of 200–443 K for Er-doped BZT-BCT ceramics. The maximum sensing sensitivity and temperature resolution were found to be 0.0044 $K^{-1}$ and 0.4 K, respectively, suggesting that Er-doped 0.5BZT-0.5BCT ferroelectric ceramic possesses potential application in optical temperature sensing [52,76].

Jiang Wu et al. reported that 0.5 mol% $Er^{3+}$ BZT-BCT ceramics, synthesized via a sol-gel synthesis route and a ceramic sintering process, possessed excellent photoluminescence performance, which is sensitive to compositional changes [52]. The morphotropic phase boundary composition exhibited the maximum photoluminescence peak intensity at 550 nm.

*4.3. Energy Storage*

In theory, the energy density J corresponds to relative permittivity and dielectric breakdown strength (BDS) according to the definition $J = 1/2\varepsilon_0\varepsilon_r E_{max}^2$. For high energy storage density, ferroelectrics are expected to possess large saturated polarization, small remnant polarization and high BDS. BZT-BCT ceramics have evoked much interest. For certain compositions, this system exhibits large permittivity of 8400, triple that of pure $BaTiO_3$ ceramics. For the $Ba(Ti, Zr)O_3$-rich compositions, this system gradually transforms to the relaxor ferroelectrics, which exhibit diffusion phase transition with broadened permittivity peaks and slim ferroelectric hysteresis loops of large maximum polarization and small remnant polarization. These characteristics may favor the high energy density applications. Venkata et al. reported a significant increase in the permittivity with relatively low dielectric losses in the Zr-rich BZT-BCT ceramics [77]. $x$BZT-BCT ($x = 0.10, 0.15, 0.20$) ceramics exhibited the electric breakdown strength as 134–170 kV/cm and high permittivity of 5200–8400. The consequent energy storage density can even reach 0.98 $J/cm^3$. Besides the pure ceramics form, the addition of glass may effectively enhance the BDS and consequently promote energy storage density [77]. Liu et al. employed glass addition $BaO$-$SrO$-$TiO_2$-$Al_2O_3$-$SiO_2$-$BaF_2$ into BZT-0.15BCT ceramics, which exhibited a large permittivity of 3458 at 25 °C under 1 kHz, slim hysteresis loop with the maximum polarization of 12.53 $\mu C/cm^2$ and a remnant polarization of 4.05 $\mu C/cm^2$ [78]. Microstructural observation indicated that the average grain size reduced significantly with increasing the glass concentration. Macroscopically, the glass ceramics exhibited diffusion phase transition with reduced peak permittivity but broad peak with relatively large permittivity of around 1000 within the room temperature region. Meanwhile the electrical breakdown strength (BDS) of the glass modified ceramics was nearly quadruple to the pure ceramics form. Energy storage performance of the glass modified ceramics, both 419.4 $kJ/m^3$ calculated from the product of permittivity and square of BDS and 192.8 $kJ/m^3$ from the integration of the hysteresis loop under the electric field of 9.6 kV/mm, showed significant superiority to that of the pure ceramic form [78]. Besides, they explored the mechanism for the enhanced energy storage property by thermal stimulated current measurements, which may reveal both polarization and charge transport process [79].

## 5. Future prospects

*5.1. Soft or Hard Modifications on BZT-BCT Ceramics*

Ferroelectrics are rarely used in a chemically pure form and doping is always employed with the goal of tailoring the properties for specific applications. Most doping effects can be categorized into "hard" and "soft" effect. "Hard" effects are obtained by acceptor doping. The defect dipoles (acceptor-oxygen vacancy) can stabilize the domain structure (no matter the volume effect or the pinning effect) [80]. As macroscopic consequences, high coercive field, low permittivity and low dielectric losses are obtained. Donor dopants result in "soft" effect, i.e. high dielectric losses, low conductivity, low coercive field and high piezoelectric coefficients [81].

Further research should focus on revealing "soft" and "hard" effects within the MPB region of the present BZT-BCT ceramics. Further, the properties of BZT-BCT ceramics can be tailored to satisfy various applications and reveal the physics behind them.

*5.2. BZT-BCT in other Forms—Single Crystal and Thin Film*

It is common sense in the piezoelectric community that the single crystal always exhibits 3–4 times higher piezoelectric effect than that of a ceramic. Besides the better properties, single crystals can also be considered as standard materials to ascertain the structure properties. However, it is difficult to grow large scale single crystals with the perovskite structures. Thus far, several methods have been employed to generate good BZT-BCT single crystal samples and have not obtained satisfactory results to date [82].

Although intensive research up to date has been inclined to bulk materials, it is certainly obvious that the films will be the focus of the future industry. Since films exhibit small volume but large geometrical flexibility, can are easy for on-chip integration, which is a prerequisite for incorporation into microelectron devices. However, the development of films still lags behind bulk counterparts. For BZT-BCT films, several techniques have been employed, such as PLD, sputtering and CSD. Hereinto, the films fabricated via PLD method exhibited $d_{33}$ of 60–140 pm/V depending on different orations [83,84]. The study of BZT-BCT film is still deficient. Further property enhancement and functionality enrichment are still the objectives to be pursued.

**Author Contributions:** Wenfeng Liu conceived and designed the manuscript, Lu Cheng wrote the manuscript, and Shengtao Li modified the manuscript.

**Funding:** This work was funded by the National Natural Science Foundation of China (51422704) and State Key Laboratory of Electrical Insulation and Power Equipment, Xi'an Jiaotong University (EIPE14113).

**Conflicts of Interest:** The authors declare no conflict of interest.

## References

1. Bernard, J.; Hans, J. Piezoelectric Ceramics. *Mullard* **1974**, *15*, 193–211.
2. Damjanovic, A. Ferroelectric, dielectric and piezoelectric properties of ferroelectric thin films and ceramics. *Rep. Prog. Phys.* **1998**, *61*, 1267–1324. [CrossRef]
3. European Commission. EU-Directive 2011/65/EU: Restriction of the use of certain hazardous substances in electrical and electronic equipment (RoHS). *Off. J. Eur. Union* **2011**, *88*, L174.
4. European Commission. Directive 2002/96/EC: Waste electrical and electronic equipment (WEEE). *Off. J. Eur. Union* **2003**, *24*, L37.
5. Rodel, J.; Jo, W.; Seifert, K.T.P.; Anton, E.; Granzow, T.; Damjanovic, D. Perspective on the Development of Lead-free Piezoceramics. *J. Am. Ceram. Soc.* **2009**, *92*, 1153–1177. [CrossRef]
6. Rubio-Marcos, F.; Ochoa, P.; Fernandez, J. Sintering and properties of lead-free (K, Na, Li) (Nb, Ta, Sb)O$_3$ ceramics. *J. Eur. Ceram. Soc.* **2007**, *27*, 4125–4129. [CrossRef]
7. Murakami, S.; Wang, D.; Mostaed, A.; Khesro, A.; Feteira, A.; Sinclair, D.; Fan, Z. High strain (0.4%) Bi(Mg$_{2/3}$Nb$_{1/3}$)O$_3$-BaTiO$_3$-BiFeO$_3$ lead-free piezoelectric ceramics and multilayers. *J. Am. Ceram. Soc.* **2018**, *101*, 5428–5442. [CrossRef]
8. Diez-Betriu, X.; Garcia, J.; Ostos, C.; Boya, A.; Ochoa, D.; Mestres, L. Phase transition characteristics and dielectric properties of rare-earth (La, Pr, Nd, Gd) doped Ba (Zr$_{0.09}$Ti$_{0.91}$)O$_3$ ceramics. *Mater. Chem. Phys.* **2011**, *125*, 493–499. [CrossRef]
9. Liu, W.; Ren, X. Large non-linear piezoelectric effect due to recoverable domain switching. In Proceedings of the 2009 18th IEEE International Symposium on the Applications of Ferroelectrics, Xi'an, China, 23–27 August 2009.
10. Bao, H.; Zhou, C.; Xue, D.; Gao, J.; Ren, X. A modified lead-free piezoelectric BZT–xBCT system with higher TC. *J. Phys. D Appl. Phys.* **2010**, *43*, 465401. [CrossRef]

11. Su, S.; Zuo, R.; Lu, S.; Xu, Z.; Wang, X.; Li, L. Poling dependence and stability of piezoelectric properties of Ba(Zr$_{0.2}$Ti$_{0.8}$)O$_3$-(Ba$_{0.7}$Ca$_{0.3}$)TiO$_3$ ceramics with huge piezoelectric coefficients. *Curr. Appl. Phys.* **2011**, *11*, S120–S123. [CrossRef]

12. Yang, R.; Fu, W.; Deng, X.; Tan, Z.; Zhang, Y.; Han, L. Preparation and Characterization of (Ba$_{0.88}$Ca$_{0.12}$)(Zr$_{0.12}$Ti$_{0.88}$)O$_3$ Powders and Ceramics Produced by Sol-Gel Process. *Adv. Mater. Res.* **2010**, *148–149*, 1062–1066. [CrossRef]

13. Ehmke, M.C.; Glaum, J.; Hoffman, M.; Blendell, J.E. In Situ X-ray Diffraction of Biased Ferroelastic Switching in Tetragonal Lead-free (1−x)Ba(Zr$_{0.2}$Ti$_{0.8}$)O$_3$–x(Ba$_{0.7}$Ca$_{0.3}$)TiO$_3$ Piezoelectrics. *J. Am. Ceram. Soc.* **2013**, *96*, 2913–2920. [CrossRef]

14. Gao, X.R.J.; Zhang, L.; Xue, D.; Kimoto, T.; Song, M.; Zhong, L. Symmetry determination on Pb-free piezoceramic 0.5Ba(Zr$_{0.2}$Ti$_{0.8}$)O$_3$-0.5(Ba$_{0.7}$Ca$_{0.3}$)TiO$_3$ using convergent beam electron diffraction method. *J. Appl. Phys.* **2014**, *115*, 054108. [CrossRef]

15. Keeble, D.S.; Benabdallah, F.; Thomas, P.; Maglione, M.; Kreisel, J. Revised structural phase diagram of (Ba$_{0.7}$Ca$_{0.3}$TiO$_3$)-(BaZr$_{0.2}$Ti$_{0.8}$O$_3$). *Appl. Phys. Lett.* **2013**, *102*, 092903. [CrossRef]

16. Woodward, D.I.; Dittmer, R.; Jo, W.; Walker, D.; Keeble, D.S.; Dale, M.W.; Rodel, J. Investigation of the Depolarisation Transition in Bi-Based Relaxor Ferroelectrics. *J. Appl. Phys.* **2014**, *115*, 114109. [CrossRef]

17. Dwivedi, A.; Qu, W.; Randall, C.A.; Damjanovic, D. Preparation and Characterization of High-Temperature Ferroelectric xBi(Mg$_{1/2}$Ti$_{1/2}$)O$_3$-yBi(Zn$_{1/2}$Ti$_{1/2}$)O$_3$–zPbTiO$_3$ Perovskite Ternary Solid Solution. *J. Am. Ceram. Soc.* **2011**, *94*, 4371–4375. [CrossRef]

18. Rossetti, G.; Khachaturyan, A.; Akcay, G.; Ni, Y. Ferroelectric Solid Solutions with Morphotropic Boundaries: Vanishing Polarization Anisotropy, Adaptive, Polar Glass, and Two-Phase States. *J. Appl. Phys.* **2008**, *103*, 114113–114115. [CrossRef]

19. Rossetti, G.A.; Khachaturyan, A. Inherent nanoscale structural instabilities near morphotropic boundaries in ferroelectric solid solutions. *Appl. Phys. Lett.* **2007**, *91*, 072909. [CrossRef]

20. Acosta, M.; Novak, N.; Jo, W.; Rödel, J. Relationship between electromechanical properties and phase diagram in the Ba(Zr$_{0.2}$Ti$_{0.8}$)O$_3$-x(Ba$_{0.7}$Ca$_{0.3}$)TiO$_3$ lead-free piezoceramic. *Acta Mater.* **2014**, *80*, 48–55. [CrossRef]

21. Yang, T.; Ke, X.; Wang, Y. Mechanisms Responsible for the Large Piezoelectricity at the Tetragonal-Orthorhombic Phase Boundary of (1-x)BaZr$_{0.2}$Ti$_{0.8}$O$_3$-xBa$_{0.7}$Ca$_{0.3}$TiO$_3$ System. *Sci. Rep.* **2016**, *6*, 33392. [CrossRef] [PubMed]

22. Wang, P.; Li, Y.; Lu, Y. Enhanced piezoelectric properties of (Ba$_{0.85}$Ca$_{0.15}$)(Ti$_{0.9}$Zr$_{0.1}$)O$_3$ lead-free ceramics by optimizing calcination and sintering temperature. *J. Eur. Ceram. Soc.* **2012**, *31*, 2005–2012. [CrossRef]

23. Liu, W.; Ren, X. Large piezoelectric effect in Pb-free ceramics. *Phys. Rev. Lett.* **2009**, *103*, 1–4. [CrossRef]

24. Tian, Y.; Wei, L.; Chao, X.; Liu, Z.; Yang, Z. Phase Transition Behavior and Large Piezoelectricity Near the Morphotropic Phase Boundary of Lead-Free (Ba$_{0.85}$Ca$_{0.15}$)(Zr$_{0.1}$Ti$_{0.9}$)O$_3$ Ceramics. *Am. Ceram. Soc.* **2013**, *96*, 496–502.

25. Xue, D.; Zhou, Y.; Bao, H.; Zhou, C.; Gao, J. Elastic, piezoelectric, and dielectric properties of Ba(Zr$_{0.2}$Ti$_{0.8}$)O$_3$-50(Ba$_{0.7}$Ca$_{0.3}$)TiO$_3$ Pb-free ceramic at the morphotropic phase boundary. *J. Appl. Phys.* **2011**, *109*, 054110. [CrossRef]

26. Shin, S.; Kim, J.; Koh, J. Piezoelectric properties of (1-x)BZT-xBCT system for energy harvesting applications. *J. Eur. Ceram. Soc.* **2018**, *38*, 4395–4403. [CrossRef]

27. Zhao, Z.; Li, X.; Ji, H.; Dai, Y.; Li, T. Microstructure and electrical properties in Zn-doped Ba$_{0.85}$Ca$_{0.15}$Ti$_{0.9}$Zr$_{0.1}$O$_3$ piezoelectric ceramics. *J. Alloys Compd.* **2015**, *637*, 291–296. [CrossRef]

28. Cui, Y.; Liu, X.; Jiang, M.; Zhao, X.; Shan, X.; Li, W.; Yuan, C.; Zhou, C. Lead-free (Ba$_{0.85}$Ca$_{0.15}$)(Ti$_{0.9}$Zr$_{0.1}$)O$_3$-CeO$_2$ piezoelectric coefficient obtained by low-temperature sintering. *Ceram. Int.* **2012**, *38*, 4761–4764. [CrossRef]

29. Huang, X.; Xing, R.; Gao, C.; Chen, Z. Influence of CeO$_2$ doping amount on property of BCTZ lead-free piezoelectric ceramics sintered at low temperature. *J. Rare Earths* **2014**, *32*, 733–737. [CrossRef]

30. Wu, Y.; Liu, X.; Zhang, Q.; Jiang, M.; Liu, Y.; Liu, X. High piezoelectric coefficient of Ba$_{0.85}$Ca$_{0.15}$Ti$_{0.9}$Zr$_{0.1}$O$_3$Sr(Cu$_{1/3}$Ta$_{2/3}$)O$_3$ ceramics. *J. Mater. Sci. Mater. Electron.* **2013**, *24*, 5199–5203. [CrossRef]

31. Chao, X.; Wang, J.; Wei, L.; Gou, R.; Yang, Z. Electrical properties and low temperature sintering of BiAlO$_3$ doped (Ba$_{0.85}$Ca$_{0.15}$)(Zr$_{0.1}$Ti$_{0.9}$)O$_3$ lead-free piezoelectric ceramics. *J. Mater. Sci.* **2015**, *26*, 7331–7340. [CrossRef]

32. Cui, Y.; Yuan, C.; Liu, X.; Zhao, X.; Shan, X. Lead-free(Ba$_{0.85}$Ca$_{0.15}$)(Ti$_{0.9}$Zr$_{0.1}$)O$_3$-Y$_2$O$_3$ ceramics with large piezoelectric coefficient obtained by low-temperature sintering. *J. Mater. Sci. Mater. Electron.* **2013**, *24*, 654–657. [CrossRef]

33. Ma, J.; Liu, X.; Jiang, M.; Yang, H.; Chen, G.; Liu, X.; Qin, L.; Luo, C. Dielectric, ferroelectric, and piezoelectric properties of Sb$_2$O$_3$-modified (Ba$_{0.85}$Ca$_{0.15}$) (Zr$_{0.1}$Ti$_{0.9}$)O$_3$ lead-free. *J. Mater. Sci. Mater. Electron.* **2014**, *25*, 992–996. [CrossRef]

34. Bai, W.F.; Li, W.; Shen, B.; Zhai, J.W. Piezoelectric and Strain Properties of Strontium-Doped BZT-BCT Lead-Free Ceramics. *High Perform. Ceram.* **2012**, *512–515*, 1385–1389. [CrossRef]

35. Wu, J.; Xiao, D.; Wu, W.; Chen, Q.; Zhu, J.; Yang, Z.; Wang, J. Role of room-temperature phase transition in the electrical properties of (Ba,Ca)(Ti,Zr)O$_3$ ceramics. *Scr. Mater.* **2011**, *65*, 771–774. [CrossRef]

36. Liu, X.; Chen, Z.; Fang, B.; Ding, J.; Zhao, X.; Xu, H.; Luo, H. Enhancing piezoelectric properties of BCZT ceramics by Sr and Sn co-doping. *J. Alloys Compd.* **2015**, *640*, 128–133. [CrossRef]

37. Chen, X.; Ruan, X.; Zhao, K.; He, X.; Zeng, J.; Li, Y.; Zheng, L.; Park, C.H.; Li, G. Low sintering temperature and high piezoelectric properties of Li-doped (Ba,Ca)(Ti,Zr)O$_3$ lead-free ceramics. *J. Alloys Compd.* **2015**, *632*, 103–109. [CrossRef]

38. Cui, Y.; Liu, X.; Jiang, M.; Wang, Y.H.; Su, Q.; Wang, H. Lead-free (Ba$_{0.7}$Ca$_{0.3}$)TiO$_3$-Ba(Zr$_{0.2}$Ti$_{0.8}$)O$_3$-xwt%CuO ceramics with high piezoelectric coefficient by low-temperature sintering. *J. Mater. Sci.* **2012**, *23*, 1342–1345. [CrossRef]

39. Liu, W.; Li, S. Effect of SiO$_2$ doping on the dielectric, ferroelectric and piezoelectric properties of (Ba$_{0.7}$Ca$_{0.3}$)(Zr$_{0.2}$Ti$_{0.8}$)O$_3$ ceramics with different sintering temperatures. *IEEE Trans. Dielectr. Electr. Insul.* **2015**, *22*, 734–738. [CrossRef]

40. Parjansri, P.; Pengpat, K.; Rujijanagul, G.; Tunkasiri, T.; Intatha, U.; Eitssayeam, S. Effect of Zn$^{2+}$ and Nb$^{5+}$ Co-Doping on Electrical Properties of BCZT Ceramics by the Seed-Induced Method. *Ferroelectrics* **2014**, *458*, 91–97. [CrossRef]

41. Xu, C.; Yao, Z.; Lu, K.; Hao, H.; Yu, Z.; Cao, M.; Liu, H. Enhanced piezoelectric properties and thermal stability in tetragonal-structured (Ba,Ca)(Zr,Ti)O$_3$ piezoelectrics substituted with trace amount of Mn. *Ceram. Int.* **2016**, *42*, 16109–16115. [CrossRef]

42. Jiang, M.; Lin, Q.; Lin, D.; Zheng, Q.; Fan, X.; Wu, X.; Sun, H.; Wan, Y.; Wu, L. Effects of MnO$_2$ and sintering temperature on microstructure, ferroelectric, and piezoelectric properties. *J. Mater. Sci.* **2013**, *48*, 1035–1041. [CrossRef]

43. Wu, J.; Wang, Z.; Zhang, B.; Zhu, J.; Xiao, D. Ba$_{0.85}$Ca$_{0.15}$Ti$_{0.90}$Zr$_{0.10}$O$_3$ Lead-free Ceramics with a Sintering Aid of MnO. *Integr. Ferroelectr.* **2013**, *141*, 89–98. [CrossRef]

44. Li, W.; Xu, Z.; Chu, R.; Fu, P.; Zang, G. Temperature Stability in Dy-Doped (Ba$_{0.99}$Ca$_{0.01}$)(Ti$_{0.98}$Zr$_{0.02}$)O$_3$ Lead-Free Ceramics with High Piezoelectric Coefficient. *J. Am. Ceram. Soc.* **2011**, *94*, 3181–3183. [CrossRef]

45. Li, W.; Xu, Z.; Chu, R.; Fu, P.; An, P. Effect of Ho doping on piezoelectric properties of BCZT ceramics. *Ceram. Int.* **2012**, *38*, 4353–4355. [CrossRef]

46. Sun, Z.; Pu, Y.; Dong, Z.; Hu, Y.; Liu, X.; Wang, P.; Ge, M. Dielectric and piezoelectric properties and PTC behavior of Ba$_{0.9}$Ca$_{0.1}$Ti$_{0.9}$Zr$_{0.1}$O$_3$−xLa ceramics prepared by hydrothermal method. *Mater. Lett.* **2014**, *118*, 1–4. [CrossRef]

47. Xia, X.; Jiang, X.; Chen, C.; Jiang, X.; Tu, N.; Chen, Y. Effects of Cr$_2$O$_3$ doping on the microstructure and electrical properties of (Ba,Ca)(Zr,Ti)O$_3$ lead-free ceramics. *Front. Mater. Sci.* **2016**, *10*, 203–210. [CrossRef]

48. Ma, J.; Liu, X.; Li, W. High piezoelectric coefficient and temperature stability of Ga$_2$O$_3$-doped. *J. Alloys Compd.* **2013**, *581*, 642–645. [CrossRef]

49. Wu, J.; Wu, W.; Xiao, D.; Wang, J.; Yang, Z.; Peng, Z.; Chen, Q.; Zhu, J. (Ba,Ca)(Ti,Zr)O$_3$-BiFeO$_3$ lead-free piezoelectric ceramics. *Curr. Appl. Phys.* **2012**, *12*, 534–538. [CrossRef]

50. Tian, Y.; Li, S.; Gong, Y.; Meng, D.; Wang, J.; Jing, Q. Effects of Er$^{3+}$–doping on dielectric and piezoelectric properties of 0.5Ba$_{0.9}$Ca$_{0.1}$TiO$_3$-0.5BaTi$_{0.88}$Zr$_{0.12}$O$_3$–0.12%La–xEr lead–free ceramics. *J. Alloys Compd.* **2017**, *692*, 797–804. [CrossRef]

51. Parjansri, P.; Intatha, U.; Eitssayeam, S. Dielectric, ferroelectric and piezoelectric properties of Nb$^{5+}$ doped BCZT ceramics. *Mater. Res. Bull.* **2015**, *65*, 61–67. [CrossRef]

52. Dai, J.; Du, P.; Xu, J.; Xu, C.; Luo, L. Piezoelectric and upconversion emission properties of Er$^{3+}$-doped 0.5Ba(Zr$_{0.2}$Ti$_{0.8}$)O$_3$-0.5(Ba$_{0.7}$Ca$_{0.3}$)TiO$_3$ ceramic. *J. Rare Earths* **2015**, *33*, 391–396. [CrossRef]

53. Han, C.; Wu, J.; Pu, C.; Qiao, S.; Wu, B.; Zhu, J.; Xiao, D. High piezoelectric coefficient of $Pr_2O_3$-doped $Ba_{0.85}Ca_{0.15}Ti_{0.90}Zr_{0.10}O_3$ ceramics. *Ceram. Int.* **2012**, *38*, 6359–6363. [CrossRef]

54. Li, Q.; Ma, W.; Ma, J.; Meng, X.; Niu, B. High piezoelectric properties of $Sm_2O_3$ doped $Ba_{0.85}Ca_{0.15}Ti_{0.90}Zr_{0.10}O_3$ ceramics. *Mater. Technol.* **2016**, *31*, 18–23. [CrossRef]

55. Tian, Y.; Gong, Y.; Meng, D.; Cao, S. Structure and electrical properties of $Ir^{4+}$-doped $0.5Ba_{0.9}Ca_{0.1}TiO_3$-$0.5BaTi_{0.88}Zr_{0.12}O_3$–0.12%La ceramics via a modified Pechini method. *Mater. Lett.* **2015**, *153*, 44–46. [CrossRef]

56. Zhou, C.; Liu, W.; Xue, D.; Ren, X.; Bao, H. Triple-point-type morphotropic phase boundary based large piezoelectricPb-freematerial—$Ba(Ti_{0.8}Hf_{0.2})O_3$-$(Ba_{0.7}Ca_{0.3})TiO_3$. *Appl. Phys. Lett.* **2012**, *100*, 222910. [CrossRef]

57. Xue, D.; Zhou, Y.; Bao, H.; Gao, J.; Zhou, C.; Ren, X. Large piezoelectric effect in Pb-free $Ba(Ti,Sn)O_3$-x(Ba,Ca)$TiO_3$ ceramics. *Appl. Phys. Lett.* **2011**, *99*, 3–6. [CrossRef]

58. Parjansri, P.; Intatha, U.; Eitssayeam, S.; Pengpat, K.; Rujijanagul, G.; Tunkasiri, T. Effects of Mn and Sr Doping on the Electrical Properties of Lead-Free 0.92BCZT-0.08BZT Ceramics. *Integr. Ferroelectr.* **2012**, *139*, 75–82. [CrossRef]

59. Liu, C.; Liu, X.; Wang, D.; Chen, Z.; Fang, B.; Ding, J. High piezoelectric coefficient of $Ba^{0.85}Ca_{0.15}Ti_{0.9}Zr_{0.1}O_3$-$Sr(Cu1/3Ta2/3)O_3$ ceramics. *J. Mater. Sci. Mater. Electron.* **2014**, *40*, 9881–9887.

60. Kalyani, A.K.; Krishnan, H.; Sen, A.; Senyshyn, A.; Ranjan, R. Polarization switching and high piezoelectric response in Sn-modified $BaTiO_3$. *Phys. Rev. B Condens. Matter Mater. Phys.* **2015**, *91*, 1–13. [CrossRef]

61. Frey, M.H.; Payne, D.A. Grain-Size Effect on Structure and Phase Transformation for Barium Titanate. *Phys. Rev. B* **1996**, *54*, 3158–3167. [CrossRef]

62. Hao, J.; Bai, W.; Li, W.; Zhai, J. Correlation Between the Microstructure and Electrical Properties in High-Performance $(Ba_{0.85}Ca_{0.15})(Zr_{0.1}Ti_{0.9})O_3$ Lead-Free Piezoelectric Ceramics. *J. Am. Ceram. Soc.* **2012**, *95*, 1998–2006. [CrossRef]

63. Brajesh, K.; Kalyani, A.; Ranjan, R. Ferroelectric instabilities and enhanced piezoelectric response in Ce modified $BaTiO_3$ lead-free ceramics. *Appl. Phys. Lett.* **2015**, *106*, 12907. [CrossRef]

64. Ang, C.; Yu, Z.; Jing, Z.; Guo, R.; Bhalla, A.; Cross, L. Piezoelectric and electrostrictive strain behavior of Ce-doped $BaTiO_3$ ceramics. *Appl. Phys. Lett.* **2002**, *80*, 3424–3426. [CrossRef]

65. Liu, W.; Zhao, D.; Li, S. Doping effect of $SiO_2/CeO_2$ on the dielectric, ferroelectric and piezoelectric properties of $(BaO_{0.7}CaO_{0.3})(ZrO_{0.2}TiO_{0.8})O_3$ ceramics. *Ferroelectrics* **2014**, *496*, 23–35.

66. Shin, S.-H.; Han, J.-D.; Yoo, J. Piezoelectric and dielectric properties of $B_2O_3$-added $(Ba_{0.85}Ca_{0.15})(Ti_{0.915}Zr_{0.085})O_3$ ceramics sintered at low temperature. *Mater. Lett.* **2015**, *154*, 120–123. [CrossRef]

67. Rubiomarcos, F.; Marchet, P.; Vendrell, X.; Romero, J.J.; Remondiere, F.; Mestres, L.; Fernandez, J.F. Effect of MnO doping on the structure, microstructure and electrical properties of the $(K,Na,Li)(Nb,Ta,Sb)O_3$ lead-free piezoceramics. *J. Alloys Compd.* **2011**, *509*, 8804–8811. [CrossRef]

68. Damjanovic, D. A morphotropic phase boundary system based on polarization rotation and polarization extension. *Appl. Phys. Lett.* **2010**, *97*, 062906. [CrossRef]

69. Fu, H.; Cohen, R.E. Polarization rotation mechanism for ultrahigh electromechanical response in single-crystal piezoelectrics. *Nature* **2000**, *403*, 281–283. [CrossRef] [PubMed]

70. Qian, X.S.; Ye, H.J.; Zhang, Y.T.; Gu, H.; Li, X.; Randall, C.; Zhang, Q. Giant electrocaloric response over a broad temperature range in modified $BaTiO_3$ ceramics. *Adv. Funct. Mater.* **2014**, *24*, 1300–1305. [CrossRef]

71. Sanlialp, M.; Shvartsman, V.V.; Acosta, M.; Dkhil, B.; Lupascu, D.C. Strong electrocaloric effect in lead-free $0.65Ba(Zr_{0.2}Ti_{0.8})O_3$-$0.35(Ba_{0.7}Ca_{0.3})TiO_3$ ceramics obtained by direct measurements. *Appl. Phys. Lett.* **2015**, *106*, 062901. [CrossRef]

72. Singha, G.; Tiwari, V.S.; Gupta, P.K. Electro-caloric effect in $(Ba_{1-x}Ca_x)(Zr_{0.05}Ti_{0.95})O_3$: A lead-free ferroelectric material. *Appl. Phys. Lett.* **2013**, *103*, 202903. [CrossRef]

73. Wang, J.; Yang, T.; Chen, S.; Li, G.; Zhang, Q. Nonadiabatic direct measurement electrocaloric effect in lead-free $Ba,Ca(Zr,Ti)O_3$ ceramics. *J. Alloy Compd.* **2013**, *550*, 561–563. [CrossRef]

74. Wang, X.; Xu, C.N.; Yamada, H.; Nishikubo, K.; Zheng, X.G. Electro-mechano-optical conversions in $Pr^{3+}$-doped $BaTiO_3$-$CaTiO_3$ ceramics. *Adv. Mater.* **2005**, *17*, 1254–1258. [CrossRef]

75. Zou, H.; Peng, D.; Wu, G.; Wang, X.; Bao, D.; Li, J.; Li, Y.; Yao, X. Polarization-induced enhancement of photoluminescence in $Pr^{3+}$ doped ferroelectric diphase $BaTiO_3$-$CaTiO_3$ ceramics. *J. Appl. Phys.* **2013**, *114*, 5–10. [CrossRef]

76. Ba, Z.O. Optical temperature sensor based on upconversion emission in Er-doped ferroelectric $0.5Ba(Zr_{0.2}Ti_{0.8})O_3$-$0.5(Ba_{0.7}Ca_{0.3})TiO_3$ ceramic. *J. Appl. Phys.* **2014**, *5*, 2–6.

77. Puli, V.S.; Pradhan, D.K.; Chrisey, D.B.; Tomozawa, M.; Sharma, G.L.; Scott, J.F.; Katiyar, R.S. Structure, dielectric, ferroelectric, and energy density properties of (1-x) BZT–xBCT ceramic capacitors for energy storage applications. *J. Mater. Sci.* **2013**, *48*, 2151–2157. [CrossRef]

78. Liu, W.; Ping, W.; Li, S. Enhanced energy storage properties of $Ba(Zr_{0.2}Ti_{0.8})O_3$-$0.15(Ba_{0.7}Ca_{0.3})TiO_3$ ceramics with $BaO$-$SrO$-$TiO_2$-$Al_2O_3$-$SiO_2$-$BaF_2$ glass addition. *Energy Technol.* **2016**, *5*, 1423–1428. [CrossRef]

79. Liu, W.; Ping, W.; Li, S. Enhanced energy storage property in glass-added $Ba(Zr_{0.2}Ti_{0.8})O_3$-$0.15(Ba_{0.7}Ca_{0.3})TiO_3$ ceramics and the charge relaxation. *Ceram. Int.* Accepted.

80. Vendrell, X.; García, J.; Bril, X.; Ochoa, D.; Mestres, L.; Dezanneau, G. Improving the functional properties of $(K_{0.5}Na_{0.5})NbO_3$ piezoceramics by acceptor doping. *J. Eur. Ceram. Soc.* **2015**, *35*, 125–130. [CrossRef]

81. Liu, W.; Chen, W.; Yang, L.; Zhang, L.; Wang, Y.; Zhou, C.; Li, S.; Ren, X. Ferroelectric aging effect in hybrid-doped $BaTiO_3$ ceramics and the associated large recoverable electrostrain. *Appl. Phys. Lett.* **2006**, *89*, 1–4. [CrossRef]

82. Palneedi, H.; Annapureddy, V.; Lee, H.; Choi, J.; Choi, S.; Chung, S.; Kangad, S.-J.L.; Ryu, J. Strong and anisotropic magnetoelectricity in composites of magnetostrictive Ni and solid-state grown lead-free piezoelectric BZT–BCT single crystals. *J. Asian Ceram. Soc.* **2017**, *5*, 36–41. [CrossRef]

83. Xu, J.B.; Shen, B.; Zhai, J.W. Dielectric, ferroelectric and optical properties of $BaZr_{0.2}Ti_{0.8}O_3$ thin films prepared by sol–gel-hydrothermal process. *J. Sol. Gel Sci. Technol.* **2010**, *55*, 343–347. [CrossRef]

84. Luo, B.; Wang, D.; Duan, M.; Li, S. Orientation-Dependent Piezoelectric Properties in Lead-Free Epitaxial $0.5BaZr_{0.2}Ti_{0.8}O_3$-$0.5Ba_{0.7}Ca_{0.3}TiO_3$ Thin Films. *Appl. Phys. Lett.* **2013**, *103*, 385–391. [CrossRef]

*Article*

# Dielectric Relaxor and Conductivity Mechanism in Fe-Substituted PMN-32PT Ferroelectric Crystal

**Xiaojuan Li** [1,2,*], **Xing Fan** [1,2], **Zengzhe Xi** [1,2], **Peng Liu** [3], **Wei Long** [1,2], **Pinyang Fang** [1,2], **Feifei Guo** [1,2] **and Ruihua Nan** [1,2]

1   School of Materials and Chemical Engineering, Xi'an Technological University, Xi'an 710032, China; fanxingabcd@163.com (X.F.); zzhxi@xatu.edu.cn (Z.X.); longwei@xatu.edu.cn (W.L.); fpy_2000@163.com (P.F.); guofeifei19850106@163.com (F.G.); nanrh@xatu.edu.cn (R.N.)
2   Shaanxi Key Laboratory of Photoelectric Functional Materials and Devices, Xi'an 710032, China
3   School of Physics and Information Technology, Shaanxi Normal University, Xi'an 710062, China; liupeng@snnu.edu.cn
*   Correspondence: lixiaojuan28@163.com; Tel.: +86-29-8617-3324

Received: 15 March 2019; Accepted: 4 May 2019; Published: 7 May 2019

**Abstract:** Fe-substituted PMN-32PT relaxor ferroelectric crystals were grown by a high-temperature flux method. The effects of charged defects on the dielectric relaxor and conductivity mechanism were discussed in detail. The Fe-substituted PMN-32PT crystal showed a high coercive field ($E_c$ = 765 V/mm), due to domain wall-pinning, induced by charged defect dipoles. Three dielectric anomaly peaks were observed, and the two dielectric relaxation peaks at low temperature were associated with the diffusion phase transition, while the high temperature one resulted from the short-range hopping of oxygen vacancies. At temperature T $\leq$ 150 °C, the dominating conduction carriers were electrons coming from the first ionization of oxygen vacancies. For the temperature range from 200 to 500 °C, the conductivity was composed of the bulk and interface between sample and electrode, and the oxygen vacancies were suggested to be the conduction mechanism. Above 550 °C, the trapped electrons from the $Ti^{3+}$ center were excited and played a major role in electrical conduction. Our results are helpful for better understanding the relationship between dielectric relaxation and the conduction mechanism.

**Keywords:** PMN-32PT single crystal; acceptor doping; charged defects; dielectric relaxation; electrical conduction

---

## 1. Introduction

Relaxor-PbTiO$_3$ ferroelectric single crystals with morphotropic phase boundary (MPB), such as $(1-x)\text{Pb}(\text{Mg}_{1/3}\text{Nb}_{2/3})\text{O}_3\text{-}x\text{PbTiO}_3$ (PMN-$x$PT) and $(1-x)\text{Pb}(\text{Zn}_{1/3}\text{Nb}_{2/3})\text{O}_3\text{-}x\text{PbTiO}_3$ (PZN-$x$PT), have attracted much attention due to their superior electric properties (piezoelectric coefficient $d_{33} \approx 2500$ pC/N, permittivity $\varepsilon \approx 5000$–7000, and strain S $\approx$ 1.7%) [1–3]. Recently, studies have shown that the ultrahigh electrical properties in relaxation ferroelectrics might derive from the local structure heterogeneity [3–6]. For example, Li et al. [5] found the contribution of polar nanoregions to the room-temperature dielectric and piezoelectric properties was up to 80% in relaxor-PT crystals. This opinion was further certified in PMN-29PT piezoelectric ceramics. By introducing the rare-earth element, samarium, to change the local structure, the piezoelectric coefficient $d_{33}$ of PMN-29PT ceramics was significantly increased from 300 to 1500 pC/N [7]. This discovery provided a new insight into improving material properties.

Heterogeneous ionic substitution was widely applied in relaxor-PT ferroelectrics to regulate their properties. It is well known that the PMN-PT is a typical "soft" piezoelectric material with a low coercive field ($E_c \approx 200$ V/mm), which greatly limits their application in high-power equipment [8,9]. In order to overcome their "softness", acceptor ionic substitution is considered an effective way to make the material "harden". For example, acceptor doping $Fe^{3+}$ and $Mn^{2+}$ ions substituted for $Ti^{4+}$ leads to a significant increase in the coercive field for PMN-$x$PT systems [9,10]. Unfortunately, the permittivity was decreased. Recent reports declared that the acceptor doping could improve not only the pyroelectric coefficient but also the mechanical quality factor of ferroelectric crystals [11,12]. The variations of electrical properties were attributed to the formation of charged defects, such as oxygen vacancies, A-site and/or B-site cation defects, electrons, and holes etc., due to the introduction of foreign ions [9–12].

The presence of charged defects is bound to change the local structure of ferroelectric material, resulting in the changes of dielectric relaxation [5,7,12–14]. Considerable studies have been performed in ferroelectric ceramics, and several possible physical mechanisms have been proposed, such as: (i) A defects dipole model related to oxygen vacancy [12–15]; (ii) Maxwell–Wagner (MW) model associated with charge of interface [16,17]; and (iii) competition between dielectric relaxation and conduction relaxation [13,18]. Compared with the ferroelectric ceramics, the effects of charged defects on the dielectric relaxation of PMN-xPT ferroelectric single crystals, especially in the conductivity mechanism, are seldom reported [19–21]. In this work, the piezoelectric, ferroelectric, and dielectric properties of Fe-substituted PMN-32PT single crystals grown by the high temperature flux method were investigated. The effects of charged defects on dielectric relaxation and conduction mechanisms of the crystals were analyzed systematically, based on complex impedance spectroscopy. Our research will be helpful for better understanding the relationship between charged defects and electrical properties in PMN-32PT ferroelectric single crystals.

## 2. Experiment

1.7 mol% Fe-substituted PMN-32PT ferroelectric single crystals were grown by the flux method. The detailed process can be found in reference [22]. A [001]-oriented crystal plate with 3 mm × 3 mm × 0.5 mm was cut and polished, and then silver paste was daubed on two sides and fired as electrodes at 750 °C for 30 minutes. A TF2000 ferroelectric test system was used to obtain the hysteresis loop at room temperature. The sample was poled under a 1 kV/mm DC electric field for 5 min at room temperature, and the piezoelectric coefficient ($d_{33}$ = 997 pC/N) was obtained by using the quasi-static meter (ZJ-6A, Institute of Acoustics, Chinese Academy of Sciences). The dielectric properties were investigated using a precision LCR meter (Agilent E4980A; Santa Rosa, CA, USA) at the temperature range of 30 to 500 °C, corresponding to the frequency range of 100 Hz to 1 MHz. The complex impedance spectroscopy and conductivity were measured using a broadband dielectric spectrometer (BDS40, Novocontrol GmbH, Montabaur, Germany) between 0.1 Hz and 3 MHz.

## 3. Results and Discussion

Figure 1 shows the powder XRD and Energy Dispersive System (EDS, see inset) of the Fe-substituted PMN-32PT crystal. X-ray powder diffraction pattern of Figure 1 contains only reflections typical for perovskite structure, indicating the Fe ions might enter the crystal. The EDS spectrum further indicates the presence of Fe element in the sample.

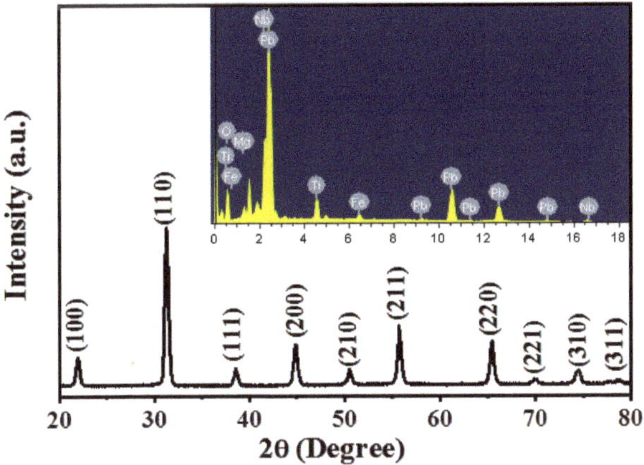

**Figure 1.** Powder XRD of Fe-substituted PMN-32PT crystal, and Energy Dispersive System (EDS) spectrum (inset).

Figure 2 shows the hysteresis loop of Fe-substituted PMN-32PT crystal at room temperature. It can be seen that the coercive field $E_c$ is 765 V/mm, which is almost three times larger than that of the pure PMN-PT crystal [8–10]. This result indicates that Fe ions effectively make PMN-$x$PT crystal "harden". One can notice that the center of the hysteresis loop shifts toward the right, leading to coercive field $E_{c-}$ = 765 V/mm and $E_{c+}$ = 721 V/mm, which might be related to the internal bias induced by charged defects in the crystal [23].

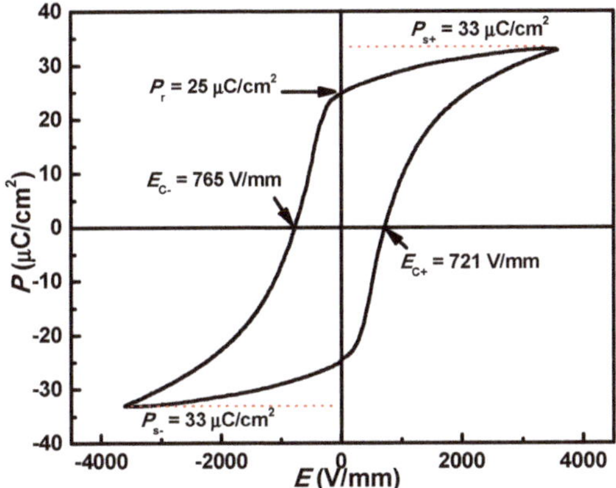

**Figure 2.** P–E loop of Fe-substituted PMN-32PT ferroelectric crystal at room temperature.

The B-site of PMN-$x$PT crystal is occupied unsystematically by $Nb^{5+}$, $Mg^{2+}$, and $Ti^{4+}$ ions. According to principles of defect chemistry, $Fe^{3+}$ ($r_{Fe^{3+}} = 0.64$Å) substitute for $Ti^{4+}$ ($r_{Ti^{4+}} = 0.61$Å) and result in the presence of charged defects. The reaction can be expressed as Equation (1):

$$Fe_2O_3 \xrightarrow{2TiO_2} 2Fe'_{Ti} + 3O_O^\times + V_O^{\bullet\bullet} \tag{1}$$

In addition, during the process of high temperature sintering, oxygen vacancies and lead vacancies are created due to their volatilization, as shown in Equations (2) and (3).

$$O_O^\times \rightarrow \frac{1}{2}O_2(g) + 2e' + V_O^{\bullet\bullet} \tag{2}$$

$$Pb_{Pb}^\times + O_O^\times \rightarrow V_{Pb}'' + V_O^{\bullet\bullet} + PbO(g) \tag{3}$$

The defect dipoles such as $2Fe_{Ti}' - V_O^{\bullet\bullet}$ may be formed, which result in an internal bias to inhibit the movement of the domain wall, accompanied by the increase of coercive field and decrease of piezoelectric coefficient [9–12], as shown in Figure 2.

Figure 3 describes the temperature dependence of DC conductivity for Fe-substituted PMN-32PT crystals. The activation energy $E_{cond}$ was calculated based on the Arrhenius formula (Equation (4)).

$$\sigma_{dc} = \sigma_0 \exp(\frac{-E_{cond}}{k_B T}) \tag{4}$$

Where $\sigma_0$ is a constant and $k_B$ is the Boltzmann constant. Three $E_{cond}$ are obtained for different temperature ranges. At low temperature T < 200 °C, the $E_{cond}$ = 0.18 eV is very close to the first ionization energy of oxygen vacancy (0.1–0.2 eV) [15], showing that the thermal ionization electrons of oxygen vacancy play a major role for the conduction mechanism. For the 500 °C > T > 200 °C temperature range, the $E_{cond}$ increases to 0.86 eV. It was reported that the migration energy of oxygen vacancies was about 1 eV in $ABO_3$ perovskite oxides [24]. Therefore, we suggested that oxygen vacancies should be responsible for the conduction mechanism. With increasing temperature above 550 °C, the activation energy $E_{cond}$ is up to 1.2 eV. Previous research suggested that the oxygen vacancies are still the dominant carriers [19,21,25]. However, our experiment results suggest that the electrons at the interface of the sample/electrode contribute to the high temperature conduction, which will be discussed in detail below.

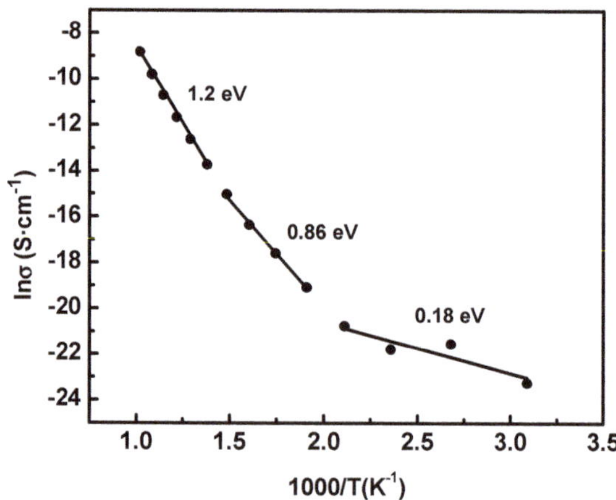


**Figure 3.** Arrhenius plots of DC conductivity as a function of temperature for Fe-substituted PMN-32PT ferroelectric crystal.

Figure 4 displays the temperature dependence of permittivity ($\varepsilon_r$) for Fe-substituted PMN-32PT single crystal at different frequencies. Two obvious dielectric anomalies are observed, corresponding to peak I and peak II. Based on previous studies [26], peak I is assigned to the rhombohedral (R)–tetragonal (T) phase transition, and peak II is from the phase transformation of tetragonal (T)–cubic (C). Fe-substituted PMN-32PT crystal shows the increased phase transition temperature of rhombohedral–tetragonal ($T_{T-R}$ = 90 °C) compared with the PMN-32PT single crystal ($T_{T-R}$ ~ 70 °C) [27], which is attributed to the domain wall-pinning induced by the internal bias field [23]. At the phase transition temperature $T_m$ = 143 °C, the permittivity $\varepsilon_r$ shows a strong frequency dependence. As frequency increases, the permittivity decreases, and peak II moves to a high temperature, indicating the relaxation property of the crystal. It's worth mentioning that there is a dielectric abnormality at high temperature 400 °C, labeled peak III. This dielectric anomaly occurs at 400 °C higher than Burns temperature $T_B$ ~ 330 °C of PMN-PT crystal [28]. Therefore, this anomaly is independent of the diffuse ferroelectric phase transition, and is known as the pseudo-dielectric relaxation [21]. When the temperature is higher than 500 °C, we observe a sharp increase in permittivity at low frequency 10 Hz, which may be associated with space charge polarization [29]. The dielectric anomaly at peak III is more obvious in the dielectric loss (see the inset of Figure 4). With the increasing of frequency, the dielectric anomaly shifts toward high temperature, indicating the dielectric relaxor.

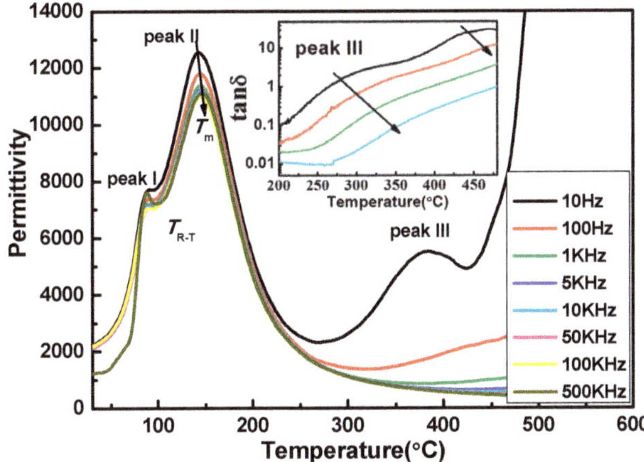


**Figure 4.** Temperature dependence of permittivity for Fe-substituted PMN-32PT ferroelectric crystal at different frequencies.

Although the high-temperature dielectric anomaly peak III can be observed in Figure 4, it is difficult to extract the accurate location of abnormal peaks, which limits our insight into the mechanism of high-temperature dielectric relaxation. Considering the relationship between dielectric modulus ($M^*$) and permittivity ($\varepsilon^*$), as in the following Equation (5), we can obtain some information about the dielectric relaxation mechanism by analysis of dielectric modulus $M^*$.

$$M^* = M' + jM'' = \frac{1}{\varepsilon^*} \tag{5}$$

Figure 5 shows the dielectric modulus imaginary $M''$ as a function of frequency ($f$) in the temperatures ranging from 200 to 700 °C. Only one peak is observed, and the peak gradually moves to high frequency with increasing temperature, which is a typical dielectric relaxation induced by thermal activation [25]. In order to analyze its physical nature, we accurately extracted the position of the relaxation peak by Gaussian fitting, and calculated the relaxation activation energy $E_{relax}$ (see inset of Figure 5), based on the Arrhenius formula (Equation (6)).

$$f_{relax} = f_0 \exp\left(\frac{-E_{relax}}{k_B T}\right) \tag{6}$$

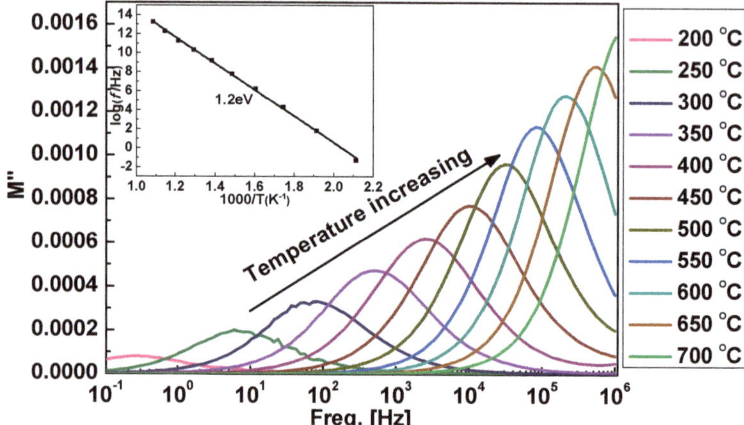



**Figure 5.** The frequency dependence of the electric modulus ($M''$) for Fe-substituted PMN-32PT ferroelectric crystal at different temperature, and relaxor activation energy ($E_{relax}$) (inset).

It can be seen that the relaxation activation energy $E_{relax}$ is 1.2 eV, indicating that the high-temperature dielectric relaxation peak III is related to the short-range hopping of oxygen vacancies [13,20,21,24]. At this point, oxygen vacancy acts as a polaron, and its hopping process can be regarded as the reorientation of the dipole, which leads to dielectric relaxation.

In order to understand the relationship between dielectric relaxation and electrical conduction, Figure 6 shows the impedance imaginary $Z''$ of Fe-substituted PMN-32PT crystal as a function of frequency ($f$) at different temperatures. Two abnormal peaks in $Z''(f)$ curves are observed in Figure 6a, corresponding to a high-frequency $P_h$ peak and a low-frequency $P_l$ peak, respectively. Usually, the high-frequency $P_h$ peak is related to the bulk, while the low-frequency $P_l$ peak is associated with interface, and they can be described by an equivalent circuit with two pairs of parallel R–C series [13,20]. With decreasing temperature from 700 to 250 °C, the position of the $P_h$ peak shifts gradually toward low frequency, indicating its relaxation characteristics. It is unfortunate that, when temperature is below 200 °C, the relaxor peak is out of this measured range and cannot be observed (see inset of Figure 6a). The intensity of the $P_h$ peak decreases gradually with the temperature increasing, showing a thermal activation. Compared with the high-frequency $P_h$ relaxation peak, the variation of the low-frequency $P_l$ peak is more complex (see Figure 6b). When the temperature is increased from 300 to 400 °C, the $P_l$ peak moves to the high frequency, accompanied by the decrease of peak intensity, which is in agreement with the $P_h$ peak. Some subtle changes are captured in the illustration. The intensity of the $P_l$ peak is higher than that of the $P_h$ peak at 300 °C, indicating that the interface's contribution to the complex impedance is greater than that of the bulk. However, the intensity of the $P_l$ peak decreases gradually with increasing temperature, and it can be hardly observed at 450 °C. It is very interesting that the $P_l$ peak rises again when the temperature is above 500 °C, but the frequency where the $Z''$ is up to maximum almost does not change with increasing temperature, except for when the peak intensity decreases. The phenomena suggest that the $P_l$ peak depends on temperature rather than frequency in the temperature range from 500 to 700 °C. Therefore, the conduction mechanism controlled by oxygen vacancy hopping is ruled out [13–17,20,21].

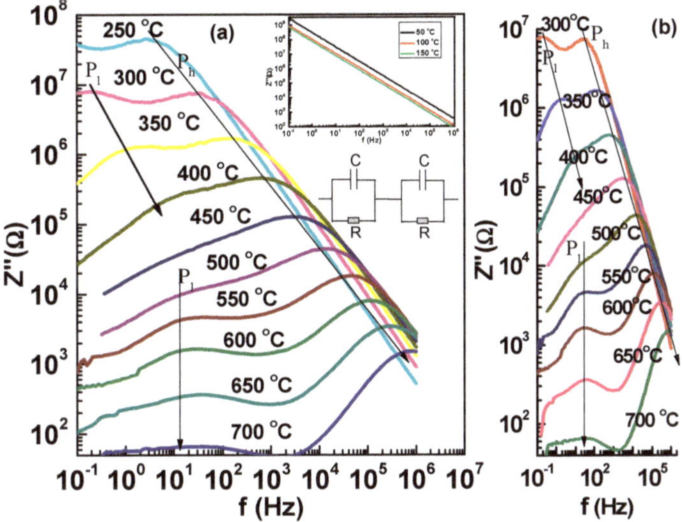

**Figure 6.** The frequency (f) dependence of the impedance imaginary part ($Z''$) for Fe-substituted PMN-32PT ferroelectric crystal at different temperatures (**a**), and the larger version (**b**). The insets show respectively the $Z''$ (f) curves and equivalent circuit.

To explore the physical mechanism of the $P_l$ and $P_h$ peaks, we extracted the position of the $P_h$ peak (250–650 °C) and $P_l$ peak (300–400 °C) from Figure 6b. Based on the Arrhenius formula (Equation (6)), the relationship between the frequency and temperature was plotted, as shown in Figure 7. It can be seen that the relaxor activation energy of $P_h$ peaks is 1.2 eV, which is in agreement with Figure 5, indicating that the dielectric relaxation and electrical conduction in bulk derive from the same physical mechanism, that is, oxygen vacancies [13,21]. For the low-frequency $P_l$ peak (300-400 °C), the relaxor activation energy is 1.28 eV, close to the migration energy of oxygen vacancy. Therefore, we suggest that the conduction mechanism of the interface arises from the oxygen vacancy in the temperature range from 300 to 400 °C [20].

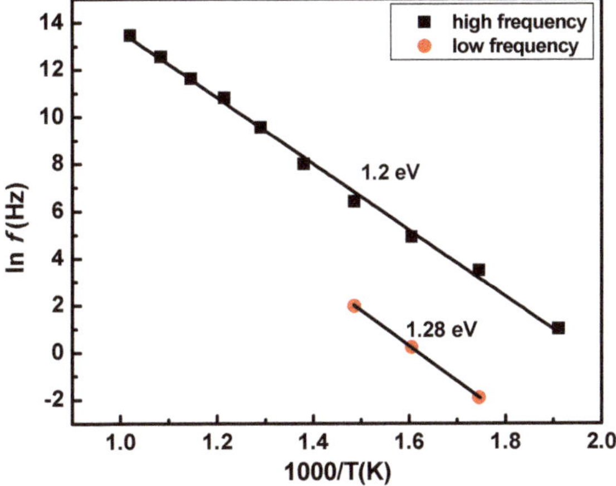


**Figure 7.** Arrhenius plots of the impedance imaginary part ($Z''$) for Fe-substituted PMN-32PT ferroelectric crystal.

For the high-temperature region (500 < T < 700 °C) in Figure 6b, as mentioned above, the $P_1$ peak shows a negative temperature resistance coefficient (NTRC) and thermal activation feature. In addition, the $P_1$ peak is independent of frequency, indicating that the interface system is a non-dispersive transport of the free charged carriers process [20]. In order to certify the view, the conductivity as a function of frequency was plotted in the temperature range 500 < T < 700 °C, as shown in Figure 8. One can see that the conductivity remains almost constant when the frequency is less than $10^3$ Hz, indicating the DC conduction. Wan et al. reported that the partial $Fe^{3+}$ might transform into $Fe^{2+}$ when the high concentration iron ions were introduced into PMN-PT crystal [11], which resulted in the formation of holes. Generally, these holes are trapped by negatively charged centers, such as $V_{Pb}''$ vacancies, and they act as carriers to form the p-type conduction at high temperature, as shown in Equation (7).

$$A^{\times} \rightarrow A' + h^{\bullet} \tag{7}$$

where $A^{\times}$ is the negatively charged center with a trapped hole. Previous study showed that the enthalpy of Equation (7) determined from the temperature dependence of the p-type conductivity is about 0.92 eV [30]. Based on Figure 2, we knew that the activation energy $E_{cond}$ was up to 1.2 eV at T > 500 °C. Therefore, we propose the holes might be excluded. However, the $Ti^{3+}$ center was calculated to be highly localized on the Ti 3d orbital and quite deep, at least 1 eV below the conduction-band edge [31]. Zhao et al. [20] reported that this activation energy corresponding to the excitation of trapped electrons from the $Ti^{3+}$ center is about $1.1 \pm 0.03$ eV in PMN-PT crystal. For the interface region between the sample and electrode, the cathodic region contains a high concentration of migrated oxygen vacancies, and they are compensated by electrons. The compensated electrons may be trapped by the $Ti^{4+}$ to form color centers, which leads to the $Ti^{4+}$ transformation into $Ti^{3+}$. Therefore, we speculate that the excitation of trapped electrons from the $Ti^{3+}$ center might dominate the high-temperature conduction mechanism [20].

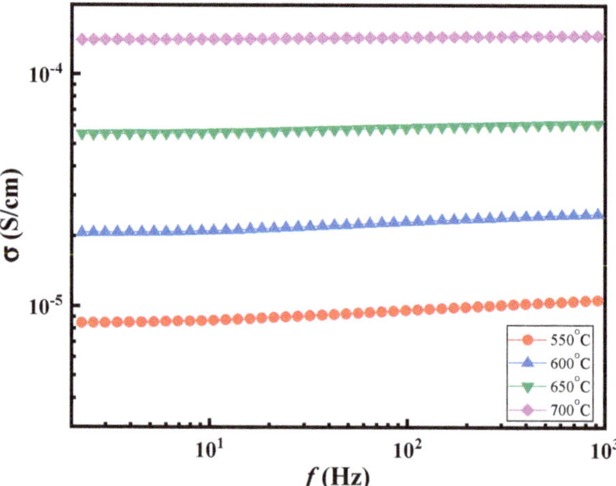


**Figure 8.** The ac conductivity of Fe-substituted PMN-32PT relaxor ferroelectric crystals as a function of frequency at different temperatures.

The impedance $Z''$ reflects the material's resistance information, while the modulus $M''$ reflects the material's capacitance information [13,18]. When the positions of $M_{max}''$ and $Z_{max}''$ are identical, they show the dominance of long-range carrier migration. Conversely, the short-range movement of carriers may play a major role in contributing to the material's properties [18]. In order to further demonstrate the carrier migration, the $M''(f)$ and $Z''(f)$ curves were contrasted at different temperatures, as shown in Figure 9. It can be clearly seen that the $Z_{max}''$ and $M_{max}''$ occur at different frequencies, as shown by the black dotted line in Figure 9a,b, which indicates that the short-range hopping of oxygen vacancies is responsible for the dielectric relaxation. With increasing temperature, the difference of frequency between $Z_{max}''$ and $M_{max}''$ decreases gradually (see Figure 9d), suggesting the dominance of oxygen vacancies' long-range migration. The long-range migration of carriers promotes conduction, resulting in the decrease of resistivity and the increase of the leakage current, as shown in Figure 6b.

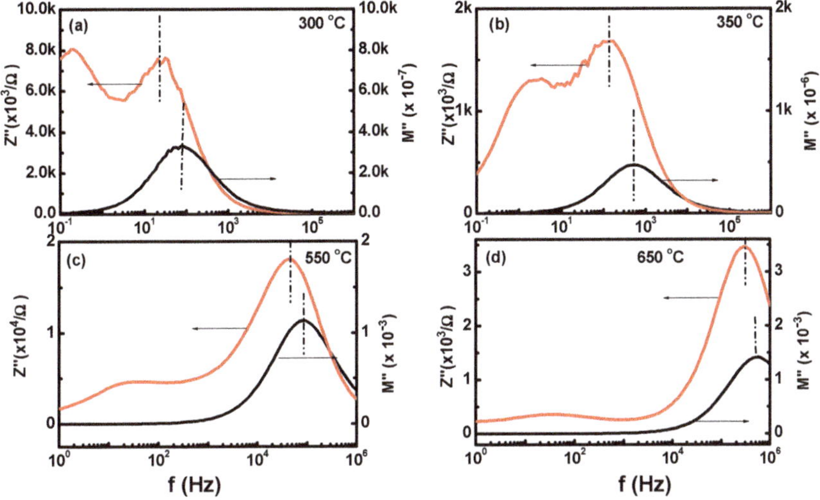

**Figure 9.** Comparison between $Z''$ and $M''$ for Fe-substituted PMN-32PT ferroelectric crystal at different temperatures (**a**) 300 °C; (**b**) 350 °C; (**c**) 550 °C and (**d**) 650 °C.

## 4. Conclusions

The dielectric, piezoelectric, ferroelectric, relaxation and electrical conduction were investigated in an Fe-substituted PMN-32PT relaxation ferroelectric single crystal, grown by the flux method. Fe ions effectively improved the coercive field $E_c$ of PMN-32PT crystal, and the $E_c$ was up to 765 V/mm, which is three times larger than that of un-doped PMN-32PT crystal. Fe-substituted PMN-32PT showed a piezoelectric coefficient $d_{33}$ = 997 pC/N. These variations of electrical properties were attributed to the domain wall-pinning by defect dipoles. The low-temperature dielectric relaxation was associated with the diffuse phase transition, while the high-temperature dielectric relaxation was attributed to the oxygen vacancies. The analysis of conductivity showed that the conduction mechanism was mainly electrons resulting from ionization of oxygen vacancies at low-temperature range T ≤ 150 °C, while the migration of oxygen vacancies dominated the conduction mechanism for moderate temperatures, 200 < T < 500 °C. As the temperature increased by 550 °C, the excitation of the trapped electrons from the $Ti^{3+}$ center below the edge of the conduction band contributed to the high-temperature conduction.

**Author Contributions:** Conceptualization, Z.X. and P.L.; methodology, X.L.; formal analysis, X.L.; investigation, X.F. and W.L.; resources, P.L.; writing—original draft preparation, X.L.; writing—review and editing, P.F.; F.G.; and R.N.; supervision, Z.X.

**Funding:** This work received financial support from the Nation Natural Science Foundation of China (Grant No. 51602242, 51772235, 11704249), and Natural Science Basic Research Plan in Shaanxi Province of China (Grant No. 2018JM5097). The authors thank the condensed matter physics laboratory of Shaanxi Normal University for the test of the broadband dielectric spectrometer.

**Conflicts of Interest:** The authors declare no conflict of interest.

## References


1.   Park, S.E.; Shrout, T.R. Ultrahigh strain and piezoelectric behavior in relaxor based ferroelectric single crystals. *J. Appl. Phys.* **1997**, *82*, 1804–1811. [CrossRef]
2.   Sun, E.; Cao, W. Relaxor-based ferroelectric single crystals: Growth, domain engineering, characterization and applications. *Prog. Mater. Sci.* **2014**, *65*, 124–210. [CrossRef] [PubMed]
3.   Qiu, C.; Liu, J.; Li, F.; Xu, Z. Thickness dependence of dielectric and piezoelectric properties for alternating current electric-field-poled relaxor-PbTiO$_3$ crystals. *J. Appl. Phys.* **2019**, *125*, 014102. [CrossRef]
4.   Xu, G.; Wen, J.; Stock, C.; Gehring, P.M. Phase instability induced by polar nanoregions in a relaxor ferroelectric system. *Nature Mater.* **2008**, *7*, 562–566. [CrossRef] [PubMed]
5.   Li, F.; Zhang, S.J.; Yang, T.; Xu, Z.; Zhang, N.; Liu, G.; Wang, J.; Cheng, Z.; Ye, Z.G.; Luo, J.; Shrout, T.R.; Chen, L.Q. The origin of ultrahigh piezoelectricity in relaxor-ferroelectric solid solution crystals. *Nat. Commun.* **2016**, *7*, 13807. [CrossRef]
6.   Manley, M.E.; Lynn, J.W.; Abernathy, D.L.; Specht, E.D.; Delaire, O.; Bishop, A.R.; Sahul, R.; Budai, J.D. Phonon localization drives polar nanoregions in a relaxor ferroelectric. *Nature Commun.* **2014**, *5*, 3683. [CrossRef]
7.   Li, F.; Lin, D.B.; Chen, Z.B.; Cheng, Z.X.; Wang, J.L.; Li, C.C.; Xu, Z.; Huang, Q.W.; Liao, X.Z.; Chen, L.Q.; Shrout, T.R.; Zhang, S.J. Ultrahigh piezoelectricity in ferroelectric ceramics by design. *Nature Mater.* **2018**, *17*, 349–354. [CrossRef]
8.   Zhang, T.F.; Tang, X.G.; Ge, P.Z.; Liu, Q.X.; Jiang, Y.P. Orientation related electrocaloric effect and dielectric phase transitions of relaxor PMN-PT single crystals. *Ceram. Int.* **2017**, *43*, 16300–16305. [CrossRef]
9.   Oh, H.T.; Joo, H.J.; Kim, M.C.; Lee, H.Y. Thickness–Dependent Properties of Undoped and Mn-doped (001) PMN–29PT[Pb(Mg$_{1/3}$Nb$_{2/3}$)O$_3$-29PbTiO$_3$] Single Crystals. *J. Korean Ceram. Soc.* **2018**, *55*, 290–298. [CrossRef]
10.  Luo, L.H.; Wang, F.F.; Tang, Y.X.; Zhu, Y.J.; Wang, J.; Luo, H.S. Fe-doped 0.71Pb(Mg$_{1/3}$Nb$_{2/3}$)O$_3$–0.29PbTiO$_3$ single crystals. *J. Phys. D: Appl. Phys.* **2008**, *41*, 205401. [CrossRef]
11.  Wan, X.M.; Chew, K.H.; Chan, H.L.W.; Choy, C.L.; Zhao, X.Y.; Luo, H.S. The effect of Fe substitution on pyroelectric properties of 0.62Pb(Mg$_{1/3}$Nb$_{2/3}$)O$_3$–0.38PbTiO$_3$ single crystals. *J. Appl. Phys.* **2005**, *97*, 064105. [CrossRef]
12.  Zheng, L.M.; Yang, L.Y.; Li, Y.R.; Lu, X.Y.; Huo, D.; Lü, W.M.; Zhang, R.; Yang, B.; Cao, W.W. Origin of Improvement in Mechanical Quality Factor in Acceptor-Doped Relaxor-Based Ferroelectric Single Crystals. *Phys. Rev. Appl.* **2018**, *9*, 064028. [CrossRef]
13.  Li, X.J.; Jing, Q.; Xi, Z.Z.; Liu, P.; Long, W.; Fang, P.Y. Dielectric relaxation and electrical conduction in (Bi$_x$Na$_{1-x}$)$_{0.94}$Ba$_{0.06}$TiO$_3$ ceramics. *J. Am. Ceram. Soc.* **2018**, *101*, 789–799. [CrossRef]
14.  Elissalde, C.; Ravez, J. Ferroelectric ceramics: defects and dielectric relaxations. *J. Mater. Chem.* **2001**, *11*, 1957–1967. [CrossRef]
15.  Chen, A.; Zhi, Y.; Cross, L.E. Oxygen-vacancy-related low-frequency dielectric relaxation and electrical conduction. *Phys. Rev. B* **2000**, *62*, 228–236.
16.  Xie, X.C.; Zhou, Z.Y.; Wang, T.Z.; Liang, R.H.; Dong, X.L. High temperature impedance properties and conduction mechanism of W$^{6+}$–doped CaBi$_4$Ti$_4$O$_{15}$ Aurivillius piezoceramics. *J. Appl. Phys.* **2018**, *124*, 204101. [CrossRef]
17.  Zhao, L.L.; Xu, R.X.; Wei, Y.X.; Han, X.; Zhai, C.X.; Zhang, Z.X.; Qi, X.F.; Cui, B.; Jonesc, J.L. Giant dielectric phenomenon of Ba$_{0.5}$Sr$_{0.5}$TiO$_3$/CaCu$_3$Ti$_4$O$_{12}$ multilayers due to interfacial polarization for capacitor applications. *J. Eur. Ceram. Soc.* **2019**, *39*, 1116–1121. [CrossRef]
18.  Wang, C.C.; Zhang, M.N.; Xu, K.B.; Wang, G.J. Origin of high-temperature relaxor-like behavior in CaCu$_3$Ti$_4$O$_{12}$. *J. Appl. Phys.* **2012**, *112*, 034109. [CrossRef]
19.  Wu, X.; Liu, L.H.; Li, X.B.; Zhao, X.Y.; Lin, D.; Luo, H.S.; Huang, Y.L. The influence of defects on ferroelectric and pyroelectric properties of Pb(Mg$_{1/3}$Nb$_{2/3}$)O$_3$–0. 28PbTiO3 single crystals. *Mater. Chem. Phys.* **2012**, *132*, 87–90. [CrossRef]
20.  Zhao, S.; Zhang, S.J.; Liu, W.; Donnelly, N.J.; Xu, Z.; Randall, C.A. Time dependent dc resistance degradation in lead-based perovskites: 0.7PbMg$_{1/3}$Nb$_{2/3}$O$_3$−0.3PbTiO$_3$. *J. Appl. Phys.* **2009**, *105*, 053705. [CrossRef]
21.  Wang, C.C.; Zhang, M.N.; Xia, W. High-Temperature Dielectric Relaxation in Pb(Mg$_{1/3}$Nb$_{2/3}$)O$_3$–PbTiO$_3$ Single Crystals. *J. Am. Ceram. Soc.* **2013**, *96*, 1521–1525. [CrossRef]
22.  He, A.G.; Xi, Z.Z.; Li, X.J.; Long, W.; Fang, P.Y.; Zhao, J.; Yu, H.B.; Kong, Y.L. Optical properties of Ho$^{3+}$- and Ho$^{3+}$/Yb$^{3+}$-modified PSN-PMN-PT crystals. *Mater. Lett.* **2018**, *219*, 64–67. [CrossRef]

23. Luo, N.N.; Zhang, S.J.; Li, Q.; Yan, Q.F.; Zhang, Y.L.; Ansella, T.; Luo, J.; Shrout, T.R. Crystallographic Dependence of Internal Bias in Domain Engineered Mn–doped Relaxor–PbTiO$_3$ Single Crystals. *J. Mater. Chem. C* **2016**, *4*, 4568–4576. [CrossRef]

24. Islam, M.S.J. Ionic transport in ABO$_3$ perovskite oxides: a computer modelling tour. *Mater. Chem.* **2000**, *10*, 1027–1038. [CrossRef]

25. Liu, X.; Fang, B.J.; Deng, J.; Deng, H.; Yan, H.; Yue, Q.W.; Chen, J.W.; Li, X.B.; Ding, J.N.; Zhao, X.Y.; Luo, H.S. Phase transition behavior and defect chemistry of [001]-oriented 0.15Pb(In$_{1/2}$Nb$_{1/2}$)O$_3$–0.57Pb(Mg$_{1/3}$Nb$_{2/3}$)O$_3$–0.28PbTiO$_3$-Mn single crystals. *J. Appl. Phys.* **2015**, *117*, 244102. [CrossRef]

26. Li, X.J.; Xi, Z.Z.; Liu, P.; Long, W.; Fang, P.Y. Stress and temperature-induced phase transitions and thermal expansion in (001)-cut PMN-31PT single crystal. *J. Alloys Compound.* **2015**, *652*, 287–291. [CrossRef]

27. Noheda, B.; Cox, D.E.; Shirane, G.; Gao, J.; Ye, Z.G. $(1-x)$PbMg$_{1/3}$Nb$_{2/3}$O$_3$–$x$PbTiO$_3$. *Phys. Rev. B* **2002**, *66*, 054104. [CrossRef]

28. Gridnev, S.A.; Glazunov, A.A.; Tsotsorin, A.N. Temperature Evolusion of the Local Order Parameter in Relaxor Ferroelectrics $(1-x)$PMN–$x$PZT. *Phys. Stat. Sol. A* **2005**, *202*, R121–R124. [CrossRef]

29. Liu, L.J.; Huang, Y.M.; Su, C.X.; Fang, L.; Wu, M.X.; Hu, C.Z.; Fan, H.Q. Space-charge relaxation and electrical conduction in K$_{0.5}$Na$_{0.5}$NbO$_3$ at high temperatures. *Appl. Phys. A* **2011**, *104*, 1047–1051. [CrossRef]

30. Raymond, M.V.; Smyth, D.M. Defects and charge transport in perovskit ferroelectrics. *J. Ph, Che. Sohds.* **1996**, *10*, 1507–1511. [CrossRef]

31. Robertson, J.; Warren, W.L.; Tuttle, B.A.; Dimos, D.; Smyth, D.M. Shallow Pb$^{3+}$ hole traps in lead zirconate titanate ferroelectrics. *Appl. Phys. Lett.* **1993**, *63*, 1519. [CrossRef]

*Article*

# Structure and Electrical Properties of Na$_{0.5}$Bi$_{0.5}$TiO$_3$ Epitaxial Films with (110) Orientation

Jianmin Song [1,2], Jie Gao [2], Suwei Zhang [3], Laihui Luo [4], Xiuhong Dai [1], Lei Zhao [1,*] and Baoting Liu [1,*]

[1]  Hebei Key Lab of Optic-Electronic Information and Materials, College of Physics Science and Technology, Hebei University, Baoding 071002, China; sjm@hebau.edu.cn (J.S.); daixiuhong@hbu.edu.cn (X.D.)

[2]  College of Science, Agriculture University of Hebei, Baoding 071001, China; jiegao1997@126.com

[3]  Center for Advanced Measurement Science, National Institute of Metrology, Beijing 100029, China; zhangsw@nim.ac.cn

[4]  College of Science, Ningbo University, Ningbo 315211, China; luolaihui@nbu.edu.cn

*  Correspondence: leizhao@hbu.edu.cn (L.Z.); btliu@hbu.edu.cn (B.L.)

Received: 17 September 2019; Accepted: 17 October 2019; Published: 25 October 2019

**Abstract:** Pt/Na$_{0.5}$Bi$_{0.5}$TiO$_3$/La$_{0.5}$Sr$_{0.5}$CoO$_3$ (Pt/NBT/LSCO) ferroelectric capacitors were fabricated on (110) SrTiO$_3$ substrate. Both NBT and LSCO films were epitaxially grown on the (110) SrTiO$_3$ substrate. It was found that the leakage current density of the Pt/NBT/LSCO capacitor is favorable to ohmic conduction behavior when the applied electric fields are lower than 60 kV/cm, and bulk-limited space charge-limited conduction takes place when the applied electric fields are higher than 60 kV/cm. The Pt/NBT/LSCO capacitor possesses good fatigue resistance and retention, as well as ferroelectric properties with P$_r$ = 35 μC/cm$^2$. The ferroelectric properties of the Pt/NBT/LSCO capacitor can be modulated by ultraviolet light. The effective polarization, ΔP, was reduced and the maximum polarization P$_{max}$ was increased for the Pt/NBT/LSCO capacitor when under ultraviolet light, which can be attributed to the increased leakage current density and non-reversible polarization P^ caused by the photo-generated carriers.

**Keywords:** NBT epitaxial film; ferroelectric properties; ultraviolet light

## 1. Introduction

Ferroelectric films are widely used in microelectromechanical systems, sensors and ferroelectric random access memory (FeRAM) due to their excellent dielectric, ferroelectric and piezoelectric properties [1–5]. The mainstay ferroelectric materials for applications have traditionally been the Pb(Zr,Ti)O$_3$ (PZT) films due to their excellent performance (large remnant polarization P$_r$ and small coercive field E$_C$). However, the use of lead gives rise to environmental concerns, which is the driving force for the development of alternative lead-free ferroelectric materials [6–8]. Na$_{0.5}$Bi$_{0.5}$TiO$_3$ (NBT) with good ferroelectric properties and high Curie temperature has been considered to be an excellent candidate to replace lead-based ferroelectric materials [9–11]. Polycrystalline NBT films with P$_r$ = 11.9 μC/cm$^2$ have been grown on Pt/Ti/SiO$_2$/Si substrates [12]. It is believed that the highly oriented ferroelectric films (especially the epitaxial films) possess higher polarization than the polycrystalline ones due to the lack of grain boundaries. Highly (111) oriented NBT film prepared on Pt/Ti/SiO$_2$/Si substrate shows a higher P$_r$ of 20.9 μC/cm$^2$ [13]. (001) and (011) oriented epitaxial NBT films fabricated on Pt-coated MgO and SrTiO$_3$ substrates by pulsed laser deposition have good dielectric and ferroelectric properties [14,15]. In addition, both P$_r$ and E$_C$ are dependent on the crystal orientation. For example, the P$_r$ and E$_C$ are 15.9 μC/cm$^2$, 12.6 μC/cm$^2$ and 126 kV/cm, 94 kV/cm for (111) and (001) oriented NBT films [14,16].

The electrode materials are very important for ferroelectric capacitors. The noble metals, such as Pt, Au and Ag, are good electrode materials due to their excellent conductivity. However, the noble metals as bottom electrodes react easily with oxygen derived from the oxide films and deteriorate the performance of the oxide films [17]. Compared to Pt, Au and Ag, the $La_{0.5}Sr_{0.5}CoO_3$ (LSCO) is low in cost and can provide an oxide/oxide (LSCO/NBT) interface that will not capture the oxygen from the oxide films [18,19]. In this work, Pt/NBT/LSCO ferroelectric capacitors were fabricated on (110)-oriented STO substrate by magnetron sputtering and pulsed laser deposition with LSCO as the bottom electrode. The microstructure and electrical properties of the Pt/NBT/LSCO ferroelectric capacitors, as well as the effect of ultraviolet light on the ferroelectric properties of these capacitors, were investigated.

## 2. Experimental

The (110) oriented Pt/NBT/LSCO/STO heterojunction was prepared by magnetron sputtering and pulsed laser deposition. Step 1: LSCO film 60 nm in thickness was deposited on (110) STO single crystal substrate by magnetron sputtering at room temperature with the following conditions: $Ar:O_2 = 3:1$, power: 30 W. Post-annealing was conducted at 550 °C in a 1 atm oxygen-flowing tube furnace. The sheet resistance of the LSCO layer was 20 $\Omega/\square$, which is quite low and would not affect the measurements. Step 2: $Na_{0.5}Bi_{0.5}TiO_3$ target with excess 10% bismuth and 10% sodium was used to compensate the loss of bismuth and sodium. NBT film with a thickness of 400 nm was deposited on the LSCO/STO heterostructure by pulsed laser deposition at 550 °C and 7.5 Pa oxygen deposition pressure. The distance between the target and substrate was 5.5 cm; the laser energy density and repetition rate were 2 J/cm$^2$ and 5 Hz, respectively. Step 3: Pt film with a thickness of 70 nm and an area of $7.85 \times 10^{-5}$ cm$^2$ was deposited by magnetron sputtering on the surface of the NBT/LSCO/STO heterostructure through a shadow mask as the top electrodes of the capacitors. The Pt/NBT/LSCO/STO heterostructure was rapidly annealed at 550 °C for 1 min under an $O_2$ atmosphere to make a better contact between the NBT and Pt.

The surface morphology of the (110) NBT film was measured by atomic force microscopy (AFM, MultiMode 8, Bruker, America). The phase structure was analyzed by X-ray diffractometer (XRD, TD-3700, Tongda, Dandong, China, Cu K$\alpha$ radiation, tube pressure 30 kV, current 20 mA). The ferroelectric properties of Pt/NBT/LSCO capacitor were tested using a ferroelectric tester (Precision LC II, Radiant, America). The leakage current of the NBT film was tested using an I-V test system (2601B, Kiethley, America ). The ultraviolet light source (CEL-HXUV300, Zhongjiaojinyuan, Beijing, China) with 365 nm (5 mW/cm$^2$) was used.

## 3. Results and Discussion

Figure 1a shows the XRD pattern of Pt/NBT/LSCO/STO heterojunction. In addition to the STO (110) peak, (110) diffraction peaks of NBT and LSCO are observed without any diffraction peaks from other directions, indicating that both NBT and LSCO films are highly (110) oriented. The full width at half maximum (FWHM) of the (110) diffraction peak for NBT film is 0.301° based on the rocking curve in Figure 1b, indicating high crystal quality. To further determine the epitaxial property of the NBT film, the phi-scan on the (100) plane of NBT film was performed as shown in Figure 1c. The two periodic diffraction peaks with similar intensity in the phi-scan further confirms that the NBT film is of good epitaxial nature. Figure 1d shows the surface topography of NBT film measured by AFM using tapping mode. It can be seen that the NBT film has a dense microstructure with a layered surface and elongated grains. The grain size is about 265 nm wide. The surface mean square roughness (RMS) is 15.5 nm, demonstrating that the NBT film has a highly crystalline quality. The elongated grain is consistent with that of the (110) oriented NBT film grown on the $SrTiO_3$ substrate reported by Bousquet [15]. In addition, the appearance of PtO in XRD can be attributed to the reaction of Pt and $O_2$, since the sample was annealed at 550 °C for 1 min in $O_2$ atmosphere.

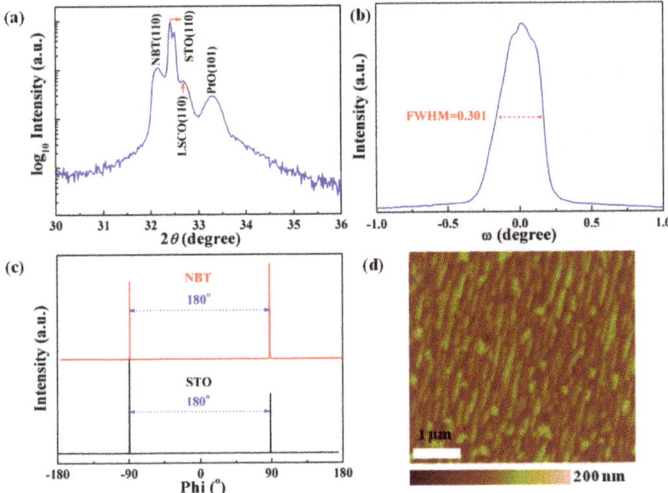

**Figure 1.** (**a**) X-ray diffraction pattern of Pt/NBT/LSCO/STO heterojunction, in which the intensity of (110) STO is normalized; (**b**) Rocking curve of (110) diffraction peak for NBT film; (**c**) Phi scan and (**d**) AFM image of NBT film.

The leakage current density has a great impact on the electrical properties of the ferroelectric capacitors. Low current density is necessary for devices. Figure 2a shows the relationship between the leakage current density J and the electric field E for Pt/NBT/LSCO ferroelectric capacitor. The leakage current density is about $4 \times 10^{-4}$ A/cm$^2$ and $2 \times 10^{-4}$ A/cm$^2$ at 250 kV/cm and $-250$ kV/cm, respectively. To further explore the conduction mechanisms of the Pt/NBT/LSCO capacitor in different electric field ranges, the leakage current curve was re-plotted as shown in Figure 2b. It was found that two mechanisms account for the leakage current characteristic of the Pt/NBT/LSCO capacitor. The log(J) and log(E) show a linear relation with a slope of 0.7 at 0~60 kV/cm, which is close to 1.0 and implies ohmic-like conduction [20,21]. There are a small number of carriers generated by thermal excitation in the NBT film, which contributes to the low J at 0~60 kV/cm. The nonlinear space-charge current-limiting mechanism is responsible for the higher J at 60~250 kV/cm. The Fermi energy is different for LSCO and Pt, which would cause a large Schottky barrier in the interfaces. A large number of electrons gathered in the electrodes under E. These electrons are activated when E is higher than the potential well and enter the NBT film to form leakage current. Thus, J increased sharply.

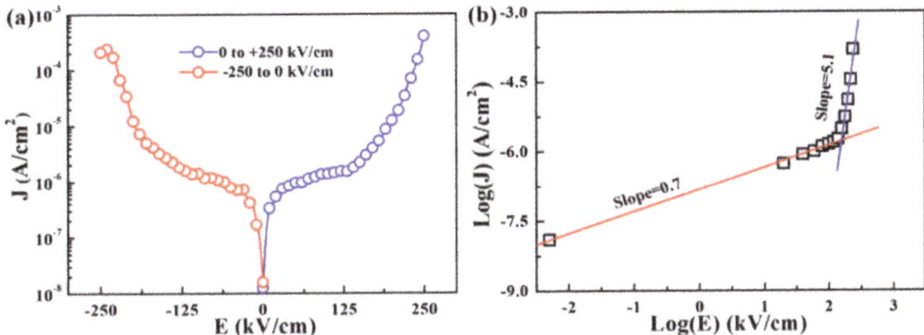

**Figure 2.** Leakage current density *vs.* applied electric fields (**a**) and log(J) *vs.* log(E) (**b**) for (110) NBT film.

Figure 3a shows the P-E loops of (110) NBT film at different E under 10 kHz. (110) NBT film shows typical ferroelectric P-E loops. Both P$_r$ and P$_{max}$ increase with E. The saturated P$_r$ of the (110) NBT film is about 35 $\mu C/cm^2$, indicating good ferroelectric properties. The $P_r$ of NBT ceramic is 38.0 $\mu C/cm^2$ [22]. In cubic structure, the angle between the (110) and (111) planes is 35.26°. In theory, $P_r = \cos(35.26°) \times 38.0$ $\mu C/cm^2 \approx 31.0$ $\mu C/cm^2$ by assuming the polarization vector is along the [111] direction. The saturated $P_r$ of (110) NBT film is 35 $\mu C/cm^2$, which is higher than that ($P_r = 31.0$ $\mu C/cm^2$) of the NBT ceramic in (110) orientation. The increased $P_r$ in (110) NBT film can be attributed to the pressure stress caused by the lattice mismatch in the NBT/LSCO interface. In the Pt/NBT/LSCO heterojunction, the in-plane lattice parameter a is 0.389 nm and 0.383 nm for NBT and LSCO based on the XRD pattern. The different lattice parameter a would lead to pressure stress and an enlarged c/a ratio (the out-of-plane lattice parameter c to the in-plane lattice parameter a) in the NBT film. The relationship between the saturation polarization $P_s$ and c/a is $(P_s)^2 \propto (c/a-1)^2$ in ferroelectric materials [23]. Thus, the pressure stress caused by the lattice mismatch will increase the polarization of the (110) NBT film.

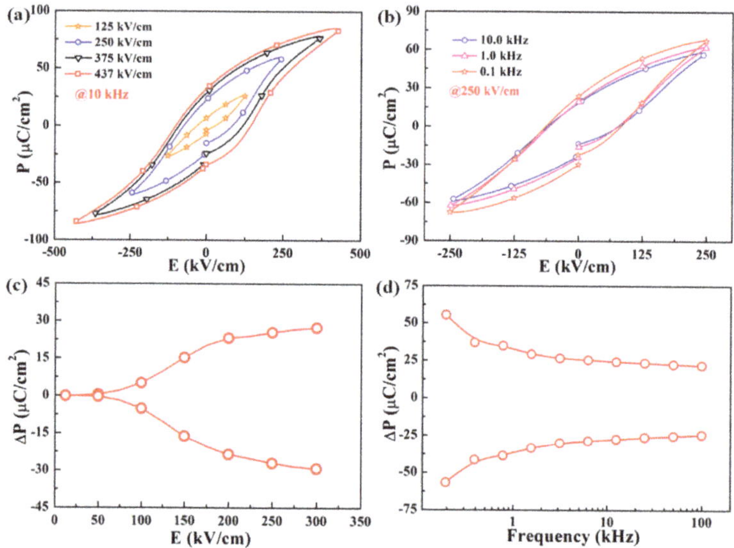

**Figure 3.** Electric field and frequency dependence of hysteresis loops (**a,b**) and ΔP (**c,d**) for (110) NBT film.

The dependence of P-E loops on the frequency under 250 kV/cm was depicted in Figure 3b. The P-E loops of (110) NBT film become weak with increasing frequency from 0.1 kHz to 10 kHz. This may be due to the fact that the domain cannot be completely reversed at high frequencies, which leads to a decrease in the polarization. The effective polarization ΔP (ΔP = P* − P^, P* is the reversible polarization, P^ is the non-reversible polarization) can remove the influence of leakage current and is an important parameter for ferroelectric memories. Figure 3c is the dependence of ΔP on E. When E is 0~50 kV/cm, ΔP remains unchanged. ΔP rapidly increases with E as E is higher than 50 kV/cm and gradually becomes saturated as E is over 175 kV/cm. At 200 kV/cm, ΔP is 20.3 $\mu C/cm^2$. The slightly increased ΔP as E > 200 kV/cm indicates that the ferroelectric domain may be completely inversed. Figure 3d is the dependence of ΔP on the frequency. The ΔP declines nonlinearly with the increase of frequency since the ferroelectric domains do not have enough time to inverse at high frequencies, which leads to reduced P*. As the frequency is higher than 1 kHz, ΔP shows weaker dependence on the frequency, indicating faster access speed.

The fatigue of the Pt/NBT/LSCO capacitor was tested at 250 kV/cm and 1 MHz, as shown in Figure 4a. No obvious degradation in $\Delta$P can be found for the LSCO bottom electrode. However, the $\Delta$P reduced to 13.5 $\mu$C/cm$^2$ from 16.5 $\mu$C/cm$^2$ after $10^{10}$ switching cycles for the Pt top electrode. These results indicate that the LSCO bottom electrode is good for fatigue resistance and the Pt top electrode would cause decreased $\Delta$P due to the reaction of Pt and O from NBT film [18,19]. The inset of Figure 4a presents the P-E loops before and after $10^{10}$ switching cycles, in which a decreased P-E loop was observed. Figure 4b is the retention of the Pt/NBT/LSCO capacitor. There is no obvious degradation in $\Delta$P for either electrode after $10^4$ s, indicating that the Pt/NBT/LSCO capacitor has good retention characteristics. In addition, the similar P-E loops before and after $10^4$ s shown in the inset of Figure 4b further confirm the good retention characteristics. It seems that the retention characteristic is independent on the electrode materials.

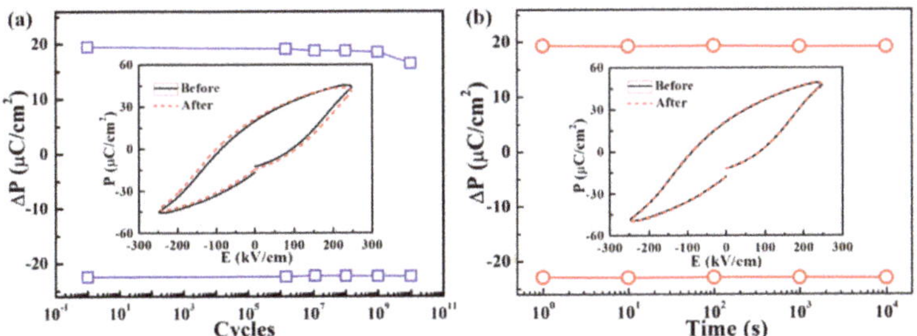

**Figure 4.** Fatigue (**a**) and retention (**b**) of Pt/NBT/LSCO capacitor. The insets are the hysteresis loops before and after $10^{10}$ switching cycles (**a**) and $10^4$ s (**b**).

To investigate the effect of light on the ferroelectric properties of the Pt/NBT/LSCO capacitor, ultraviolet light with a wavelength of 365 nm was used, since the forbidden gap of NBT film is 3.15 eV, as shown in Figure 5a. Figure 5b shows the P-E loops of the Pt/NBT/LSCO capacitor under dark and ultraviolet light. It can be seen that the P$_{max}$ is 60.6 $\mu$C/cm$^2$ and 63.8 $\mu$C/cm$^2$ under dark and ultraviolet light, respectively, indicating that the ultraviolet light can increase the polarization of the Pt/NBT/LSCO capacitor. P$_{max}$ (P$_{max}$ = 2$\Delta$P + Jt, $\Delta$P is effective polarization, J is leakage current density and t is time) can be affected by $\Delta$P and J. Under ultraviolet light, photo-generated carriers would be generated in NBT film, increasing J, and thus causing an increase in P$_{max}$. To further illustrate the effect of ultraviolet light on $\Delta$P, the dependences of $\Delta$P on E and frequency under ultraviolet light were investigated as shown in Figure 5c,d. It can be seen that the ultraviolet light leads to reduced $\Delta$P. This is attributed to the increased P$^\wedge$ caused by the ultraviolet light. The decreased $\Delta$P further confirms that the increased P$_{max}$ can be attributed to the increased J under ultraviolet light. The ultraviolet light leads to decreased $\Delta$P, but does not change the tendencies of $\Delta$P with E and frequency. Based on the analysis described above, it can be concluded that the increased P$_{max}$ in the Pt/NBT/LSCO capacitor is due to the increased J caused by the photo-generated carriers rather than the increased $\Delta$P.

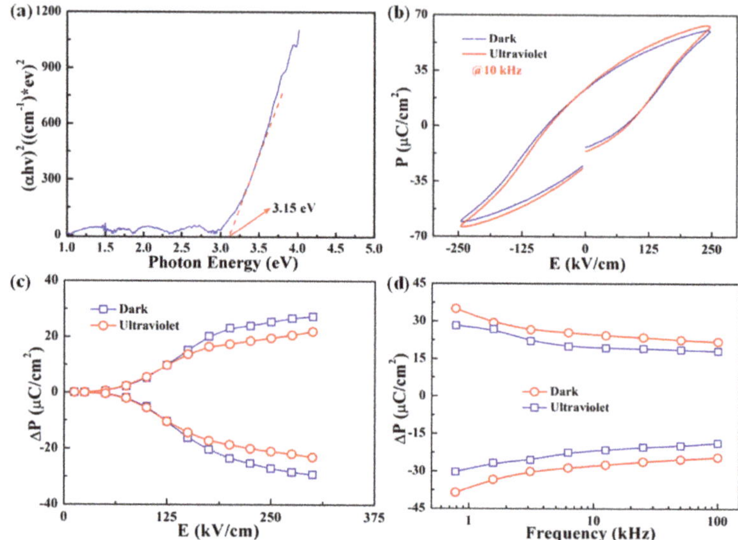

**Figure 5.** Forbidden gap (**a**), hysteresis loops (**b**), electric field-dependence (**c**) and frequency-dependence (**d**) of ΔP for Pt/NBT/LSCO capacitor under ultraviolet light.

## 4. Conclusions

The Pt/NBT/LSCO capacitor was fabricated on (110) STO substrate by magnetron sputtering and pulsed laser deposition with LSCO as the bottom electrode. The microstructure and electrical properties of the (110) oriented NBT film were investigated. It was found that the (110) oriented NBT/LSCO films were epitaxially grown on $SrTiO_3$ substrate with high crystal quality. The (110) NBT film shows good ferroelectric properties, with $P_r = 35$ μC/cm$^2$ and a small leakage current density of $4.02 \times 10^{-4}$ A/cm$^2$ at 250 kV/cm. The ohmic conduction mechanism and nonlinear space charge limiting accounted for the conduction mechanisms at 0–60 kV/cm and 60–250 kV/cm, respectively. The ΔP of the Pt/NBT/LSCO capacitor shows strong dependence on both the electric field and frequency. In addition, the Pt/NBT/LSCO capacitor processes good fatigue resistance and retention. The ultraviolet light leads to increased leakage current density and non-reversible polarization P^, and causes reduced ΔP and increased $P_{max}$. These results can provide a reference for the research and development of lead-free NBT ferroelectric storage devices.

**Author Contributions:** J.S. and J.G. contributed equally to this work in sample preparation and testing. S.Z. worked for data analysis and English. L.L. and X.D. worked for data analysis. L.Z. and B.L. worked for manuscript writing.

**Funding:** This research received no external funding.

**Acknowledgments:** This project is supported by the National Natural Science Foundation of China (Grant numbers 11374086 and 51802068), the Natural Science Foundation of Hebei Province (Grant numbers E2014201188, E2014201063 and A2018201168), the Advanced Talents Incubation Program of Hebei University (Grant numbers 801260201180 and 521000981323) and the State Key Laboratory of New Ceramic and Fine Processing Tsinghua University (Grant number KF201812).

**Conflicts of Interest:** The authors declare no conflict of interest.

## References


1. Peddigari, M.; Palneedi, H.; Hwang, G.T.; Lim, K.W.; Kim, G.Y.; Jeong, D.Y.; Ryu, J. Boosting the Recoverable Energy Density of Lead-Free Ferroelectric Ceramic Thick Films through Artificially Induced Quasi-Relaxor Behavior. *ACS Appl. Mater. Interfaces* **2018**, *10*, 20720–20727. [CrossRef]

2. Tang, Z.; Liu, Z.; Ma, J.; Fan, J.; Zhong, M.; Tang, X.G.; Lu, S.G.; Tang, M.; Gao, J. The enhanced magnetoelectric effect and piezoelectric properties in the lead-free $Bi_{3.15}Nd_{0.85}Ti_3O_{12}/La_{0.7}Ca_{0.3}MnO_3$ nano-multilayers composite thin films. *J. Alloys Compd.* **2019**, *777*, 485–491. [CrossRef]

3. Wen, Z.; Li, C.; Wu, D.; Li, A.; Ming, N. Ferroelectric-field-effect-enhanced electroresistance in metal/ferroelectric/semiconductor tunnel junctions. *Nat. Mater.* **2013**, *12*, 617–621. [CrossRef]

4. Tian, B.B.; Liu, Y.; Chen, L.F.; Wang, J.L.; Sun, S.; Shen, H.; Sun, J.L.; Yuan, G.L.; Fusil, S.; Garcia, V.; et al. Space-charge Effect on Electroresistance in Metal-Ferroelectric-Metal capacitors. *Sci. Rep.* **2015**, *5*, 18297. [CrossRef] [PubMed]

5. Abuwasib, M.; Lu, H.; Li, T.; Buragohain, P.; Lee, H.; Eom, C.B.; Gruverman, A.; Singisetti, U. Scaling of electroresistance effect in fully integrated ferroelectric tunnel junctions. *Appl. Phys. Lett.* **2016**, *108*, 152904. [CrossRef]

6. Zhang, T.; Li, W.; Cao, W.; Hou, Y.; Yu, Y.; Fei, W. Giant electrocaloric effect in PZT bilayer thin films by utilizing the electric field engineering. *Appl. Phys. Lett.* **2016**, *108*, 162902. [CrossRef]

7. Katsouras, I.; Asadi, K.; Groen, W.A.; Blom, P.W.M.; De Leeuw, D.M.; Zhao, D. Retention of intermediate polarization states in ferroelectric materials enabling memories for multi-bit data storage. *Appl. Phys. Lett.* **2016**, *108*, 232907. [CrossRef]

8. Gobeljic, D.; Shvartsman, V.V.; Belianinov, A.; Okatan, B.; Jesse, S.; Kalinin, S.V.; Groh, C.; Rödel, J.; Lupascu, D.C.; Okatan, M.; et al. Nanoscale mapping of heterogeneity of the polarization reversal in lead-free relaxor–ferroelectric ceramic composites. *Nanoscale* **2016**, *8*, 2168–2176. [CrossRef] [PubMed]

9. Smolenskii, G.A.; Isypov, V.A.; Agranovskaya, A.I.; Krainik, N.N. New ferroeleetrics complex composition. *Sov. Phys. Solid State* **1961**, *2*, 2651–2654.

10. Cao, W.; Li, W.; Feng, Y.; Bai, T.; Qiao, Y.; Hou, Y.; Zhang, T.; Yu, Y.; Fei, W. Defect dipole induced large recoverable strain and high energy-storage density in lead-free $Na_{0.5}Bi_{0.5}TiO_3$-based systems. *Appl. Phys. Lett.* **2016**, *108*, 202902. [CrossRef]

11. Balakt, A.; Shaw, C.; Zhang, Q. Enhancement of pyroelectric properties of lead-free $0.94Na_{0.5}Bi_{0.5}TiO_3$-$0.06BaTiO_3$ ceramics by La doping. *J. Eur. Ceram. Soc.* **2017**, *37*, 1459–1466. [CrossRef]

12. Zhou, Z.H.; Xue, J.M.; Li, W.Z.; Wang, J.; Zhu, H.; Miao, J.M. Leakage current and charge carriers in $(Na_{0.5}Bi_{0.5})TiO_3$thin film. *J. Phys. D Appl. Phys.* **2005**, *38*, 642–648. [CrossRef]

13. Tang, X.G.; Wang, J.; Wang, X.X.; Chan, H.L.W. Preparation and Electrical Properties of Highly (111)-Oriented $(Na_{0.5}Bi_{0.5})TiO_3$ Thin Films by a Sol−Gel Process. *Chem. Mater.* **2004**, *16*, 5293–5296. [CrossRef]

14. Bousquet, M.; Duclère, J.-R.; Champeaux, C.; Boulle, A.; Marchet, P.; Catherinot, A.; Wu, A.; Vilarinho, P.M.; Députier, S.; Guilloux-Viry, M.; et al. Macroscopic and nanoscale electrical properties of pulsed laser deposited (100) epitaxial lead-free $Na_{0.5}Bi_{0.5}TiO_3$ thin films. *J. Appl. Phys.* **2010**, *107*, 034102. [CrossRef]

15. Bousquet, M.; Duclère, J.-R.; Gautier, B.; Boulle, A.; Wu, A.; Deputier, S.; Fasquelle, D.; Rémondière, F.; Albertini, D.; Champeaux, C.; et al. Electrical properties of (110) epitaxial lead-free ferroelectric $Na_{0.5}Bi_{0.5}TiO_3$ thin films grown by pulsed laser deposition: Macroscopic and nanoscale data. *J. Appl. Phys.* **2012**, *111*, 104106. [CrossRef]

16. Duclère, J.-R.; Cibert, C.; Boulle, A.; Dorcet, V.; Marchet, P.; Champeaux, C.; Catherinot, A.; Deputier, S.; Guilloux-Viry, M. Lead-free $Na_{0.5}Bi_{0.5}TiO_3$ ferroelectric thin films grown by Pulsed Laser Deposition on epitaxial platinum bottom electrodes. *Thin Solid Films* **2008**, *517*, 592–597. [CrossRef]

17. Solbach, A.; Klemradt, U.; Schorn, P.J.; Böttger, U.; Cao, J.L.; Weirich, T.E.; Mayer, J. Probing fatigue in ferroelectric thin films with subnanometer depth resolution. *Appl. Phys. Lett.* **2007**, *91*, 72905.

18. Liu, B.T.; Chen, J.E.; Sun, J.; Wei, D.Y.; Chen, J.H.; Li, X.H.; Bian, F.; Zhou, Y.; Guo, J.X.; Zhao, Q.X.; et al. Oxygen vacancy as fatigue evidence of $La_{0.5}Sr_{0.5}CoO_3/PbZr_{0.4}Ti_{0.6}O_3/La_{0.5}Sr_{0.5}CoO_3$ capacitors. *EPL Europhys. Lett.* **2010**, *91*, 67011. [CrossRef]

19. Song, J.M.; Luo, L.H.; Dai, X.H.; Song, A.Y.; Zhou, Y.; Li, Z.N.; Liang, J.T.; Liu, B.T. Switching properties of epitaxial $La_{0.5}Sr_{0.5}CoO_3/Na_{0.5}Bi_{0.5}TiO_3/La_{0.5}Sr_{0.5}CoO_3$ ferroelectric capacitor. *RSC Adv.* **2018**, *8*, 4372–4376. [CrossRef]

20. Cheng, B.L.; Wang, C.; Zhou, Y.L.; Wang, S.Y.; Dai, S.Y.; Lu, H.B.; Chen, Z.H.; Yang, G.Z. Reduction of leakage current by Co doping in Pt/Ba$_{0.5}$Sr$_{0.5}$TiO$_3$/Nb–SrTiO$_3$ capacitor. *Appl. Phys. Lett.* **2004**, *84*, 4116.

21. Qi, Y.; Lu, C.; Zhang, Q.; Wang, L.; Chen, F.; Cheng, C.; Liu, B. Improved ferroelectric and leakage properties in sol–gel derived BiFeO$_3$/Bi$_{3.15}$Nd$_{0.85}$Ti$_3$O$_{12}$ bi-layers deposited on Pt/Ti/SiO$_2$/Si. *J. Phys. D Appl. Phys.* **2008**, *41*, 065407. [CrossRef]

22. Hiruma, Y.; Nagata, H.; Takenaka, T. Thermal depoling process and piezoelectric properties of bismuth sodium titanate ceramics. *J. Appl. Phys.* **2009**, *105*, 084112. [CrossRef]

23. Fatuzzo, E.; Merz, W.J. *Ferroelectricity.*; North-Holland Publishing Company: Amsterdam, The Netherlands, 1967.

 *crystals* 

*Article*

# Capacitance Properties in Ba$_{0.3}$Sr$_{0.7}$Zr$_{0.18}$Ti$_{0.82}$O$_3$ Thin Films on Silicon Substrate for Thin Film Capacitor Applications

Xiaoyang Chen, Taolan Mo, Binbin Huang, Yun Liu and Ping Yu *

College of Material Science and Engineering, Sichuan University, Chengdu 610064, China;
chenxy.189@gmail.com (X.C.); mtaolan@gmail.com (T.M.); huangbinbin@stu.scu.edu.cn (B.H.);
liuyun6@stu.scu.edu.cn (Y.L.)
* Correspondence: pingyu@scu.edu.cn

Received: 30 March 2020; Accepted: 17 April 2020; Published: 19 April 2020

**Abstract:** Crystalline Ba$_{0.3}$Sr$_{0.7}$Zr$_{0.18}$Ti$_{0.82}$O$_3$ (BSZT) thin film was grown on Pt(111)/Ti/SiO$_2$/Si substrate using radio frequency (RF) magnetron sputtering. Based on our best knowledge, there are few reports in the literature to prepare the perovskite BSZT thin films, especially using the RF magnetron sputtering method. The microstructure of the thin films was characterized using X-ray diffraction (XRD) and scanning electron microscopy (SEM), and capacitance properties, such as capacitance density, leakage behavior, and the temperature dependence of capacitance were investigated experimentally. The prepared perovskite BSZT film showed a low leakage current density of $7.65 \times 10^{-7}$ A/cm$^2$ at 60 V, and large breakdown strength of 4 MV/cm. In addition, the prepared BSZT thin film capacitor not only exhibits an almost linear and acceptable change ($\Delta C/C$ ~13.6%) of capacitance from room temperature to 180 °C but also a large capacitance density of 1.7 nF/mm$^2$ at 100 kHz, which show great potential for coupling and decoupling applications.

**Keywords:** BSZT thin films; capacitance properties; RF magnetron sputtering

## 1. Introduction

Recently, the increasing demand for high-density and highly integrated electronic passive components in the microelectronics industry, has greatly accelerated research on thin film capacitors with high capacitance density (capacitance per unit area) and advanced functional dielectric films with high dielectric constant and appropriate dielectric strength [1–3]. Except for high capacitance density and appropriate electric breakdown strength, temperature stability is another very important parameter for dielectric capacitor applications [4]. Many international electronic industries alliances, like the now-defunct Electronic Industries Alliance (EIA) and European Committee for Electrotechnical Standardization (CENELEC), have clear demands of different temperature stability for various capacitor applications [5]; for instance, $\Delta C/C$ over the temperature range in X7R ceramics capacitor is lower than $-15\%$ to $+15\%$. According to these standards, the ferroelectric materials which exhibit paraelectric behavior over the working temperature range, can not only have a bigger dielectric constant than the linear dielectrics, but also can satisfy some EIA Class I or EIA Class II ceramics capacitors with a high requirement of temperature stability, like high-frequency capacitors, and coupling and decoupling applications.

As one of the most studied perovskite material systems, SrTiO$_3$ (STO) thin films have been widely used in these applications due to its high dielectric constant (~120), low dielectric loss, and low curie temperature (~250 °C) [6–11]. Compared with STO films, the Ba$_{1-x}$Sr$_x$TiO$_3$ (BST) thin films have a higher dielectric constant, which has currently become very attractive for higher integration of thin film capacitors. However, electronic hopping between Ti$^{4+}$ and Ti$^{3+}$ ions makes the BST thin films have a low electric breakdown strength (hardly exceeding 0.8 MV/cm) [12,13]. For this reason, $Ba_xSr_{1-x}Zr_yTi_{1-y}O_3$ (BSZT) ceramics are thus fabricated by the substitution of $Zr^{4+}$ for $Ti^{4+}$ in BST ceramics [14]. As $Zr^{4+}$ is chemically more stable than $Ti^{4+}$ and has a larger ionic size to expand the perovskite lattice, the substitution of Ti with Zr can improve the chemical and temperature stability, and reduce the dielectric loss [15,16]. Therefore, the BSZT thin films could be a promising dielectric thin film as an alternative to the BST or STO thin films for the high-frequency capacitors, coupling and decoupling capacitors. Thus far, the perovskite BSZT thin films are prepared by using pulse laser deposition (PLD) and Sol-Gel processing [13,15–19]. Radio frequency (RF) magnetron sputtering method is widely used in the preparation of advanced functional thin films; it is especially flexible in preparing high-quality oxide thin films with a large area. Based on our best knowledge, there are few reports in the literature to prepare the perovskite BSZT thin films [13,15–19], especially using the RF magnetron sputtering method.

In this work, the perovskite $Ba_{0.3}Sr_{0.7}Zr_{0.18}Ti_{0.82}O_3$ thin films were prepared on Pt (111)/$SiO_2$/Si substrates by using RF magnetron sputtering technology. According to our previous results, the Curie temperature (*Tc*) of $Ba_{0.3}Sr_{0.7}Zr_{0.18}Ti_{0.82}O_3$ compounds is about $-102\,°C$, thus the $Ba_{0.3}Sr_{0.7}Zr_{0.18}Ti_{0.82}O_3$ thin films exhibit para-electric behavior above that temperature. Its capacitance properties were systematically characterized as a function of frequency, voltage, and temperature. In addition, its dielectric performance was compared with other reported dielectric thin-films.

## 2. Materials and Methods

The radio frequency (RF) magnetron sputtering method was used to deposit BSZT thin films on Pt(111)/$SiO_2$/Si(100) substrates using a RF magnetron sputtering device (Sky Technology Development, JGP560D, Shenyang, China). In this experiment, the oxygen and argon gas were mixed with a ratio of 1:4 and a total pressure of 2 Pa, and the growth temperature was carried at 625 °C. The BSZT powders were prepared as the target with a purity of 99%. This deposition time is 5 hours. After the sputtering process, the deposited thin films were annealed by conventional thermal annealing (CTA) at 700 °C for 180 min in the air to obtain well-crystallized grains.

The crystal phase of the BSZT thin films was measured by using X ray-diffraction (DX-2700, Dandong, China) with Cu K$\alpha$ radiation. The thickness of the BSZT thin films was investigated by field emission scanning electron microscope (JEOL, JSM-7500F, Tokyo, Japan). The top Au electrode layer was prepared with an area of 0.145 $mm^2$. The dielectric properties were measured by a multi-frequency Inductance-Capacitance-Resistance (LCR) meter (Agilent, HP4294A, Santa Clara, USA). The leakage current behavior and polarization-electric field (*P-E*) hysteresis loops of the capacitor were measured by a Radiant Precision Workstation (Radiant Technologies, Median, New York, USA). The temperature-related properties were carried on a heating probe stage (Linkam, THMS600, Surrey, UK).

## 3. Results

The X-ray diffraction (XRD) patterns of the BSZT films and BSZT ceramics are shown in Figure 1A. It is agreed that the detected XRD peaks are almost the same, which indicates the perovskite BSZT phase was clearly formed on the Si/$SiO_2$/Pt(111) substrate. The deposited films have a pure perovskite phase and no secondary phase is detected, confirming that the stable solid solution of BSZT is formed. The cross-section SEM photograph in Figure 1B shows that the prepared BSZT thin films show a dense, cracks-free, uniform microstructure, and the clear structure of the Pt/BSZT/Au thin film capacitor is visible. The thickness of the prepared BSZT films is about 870 nm.

Figure 2 shows the capacitance properties of the BSZT films as a function of frequency, voltage, and temperature. As shown in Figure 2A, the prepared BSZT films show a weak frequency dispersion and a large capacitance density of 1.7 nF/mm$^2$ at 100 kHz. It can be calculated that the dielectric constant ($\varepsilon$r) is about 170, which is obviously higher than the reported STO films (~100). Therefore, a larger capacitance density higher than that of STO thin films could be achieved. In addition, the prepared BSZT films show a very low dielectric loss of 0.01, which is lower than the reported BST thin films [20]. The results reveal that the substitution of Ti with Zr can reduce the dielectric loss. The Direct Current (DC) voltage dependence of the capacitance density of the prepared films at the frequency of 100 kHz was measured from −40 V to 40 V in Figure 2B. The prepared films can withstand 40 V DC bias voltage. It was also noticed that the capacitance decreases with the voltage. This phenomenon is very prevalent in EIA Class II ceramic capacitor. For a better description of the stability of the capacitance under high operating voltage, the relative variation ratio of capacitance was plotted as a function of DC voltage in Figure 2C. This ratio is calculated by the change of the capacitance values under different DC operating voltages to that of 0 V DC. The EIA code does not take into consideration the DC voltage dependence in the Class II ceramics capacitor. According to the CENELEC Electronic Components Committee (CECC) code, the DC voltage dependence is less than −30% for the 2C1 class ceramics capacitor [21]. Therefore, the prepared BSZT films can satisfy this standard when the rated voltage is less than 40 V. The temperature dependence of capacitance in the testing range of 30~180 °C is shown in Figure 2D. The prepared capacitance exhibits an almost linear and small change ($\Delta$C/C ~13.6%) of capacitance from 30 to 180 °C. According to the EIA code, the BSZT thin film capacitor can satisfy the requirement from most Class II ceramics with first letter code of Z.

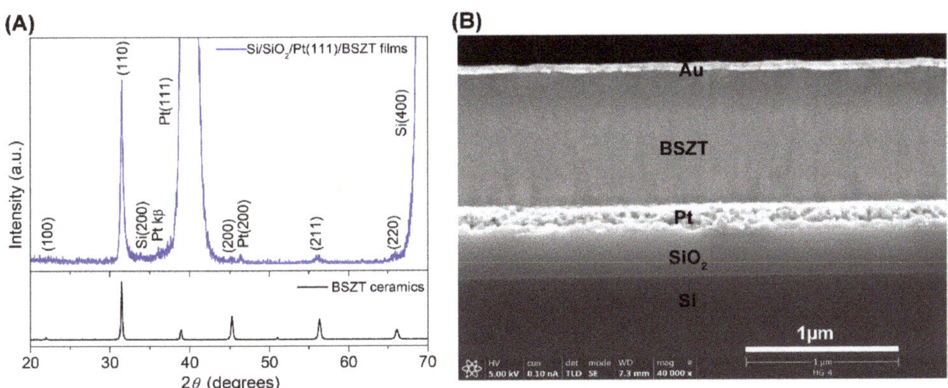


**Figure 1.** (**A**) The X ray-diffraction (XRD) patterns of the Ba$_{0.3}$Sr$_{0.7}$Zr$_{0.18}$Ti$_{0.82}$O$_3$ (BSZT) films. (**B**) The cross-section scanning electron microscopy (SEM) photographs of the Pt/BSZT/Au thin film capacitor.

Figure 3A shows the polarization electric field (*P-E*) hysteresis loop and the leakage current behavior of the BSZT thin films. The *P-E* loops of the prepared BSZT films possess good linearity and weak ferroelectric behavior, which indicate the obvious characteristics of paraelectric thin films. Figure 3B shows the leakage current density of the thin films as a function of the applied bias electrical field from −60 V to 60 V. The leakage current density of the films linearly increases with the applied bias range, which indicates good ohmic conduction in the films. More importantly, the current density is about 7.65 × 10$^{-7}$ A/cm$^2$ at 60 V, which is lower than that of BST films.

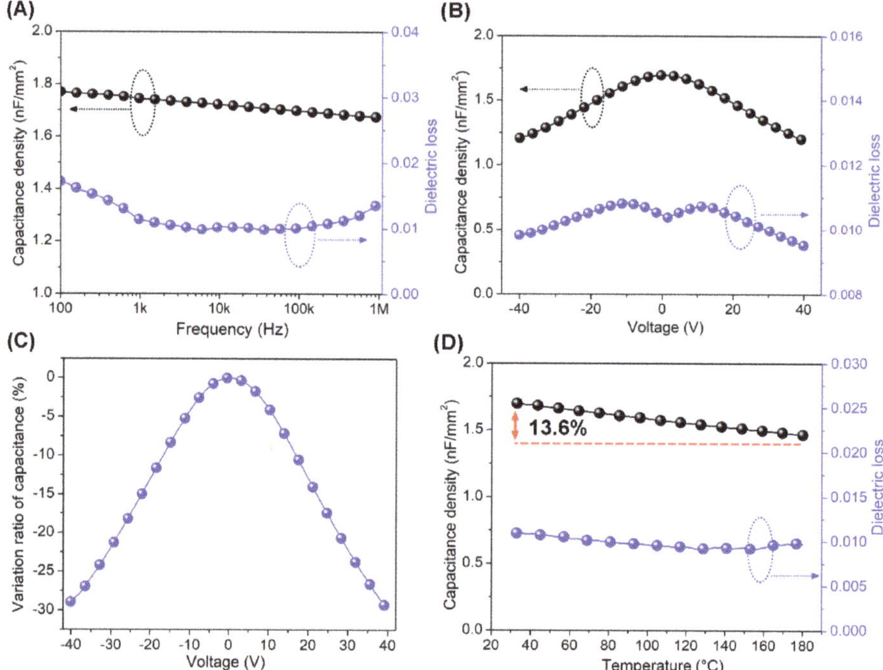

**Figure 2.** The capacitance properties of the BSZT films. (**A**) The capacitance density and dielectric loss as a function of frequency. (**B**) Direct Current (DC) voltage dependence of capacitance density and dielectric loss. (**C**) The relative variation ratio of capacitance density as a function of DC voltage. (**D**) The capacitance density as a function of testing temperature in the range of 30~180 °C.

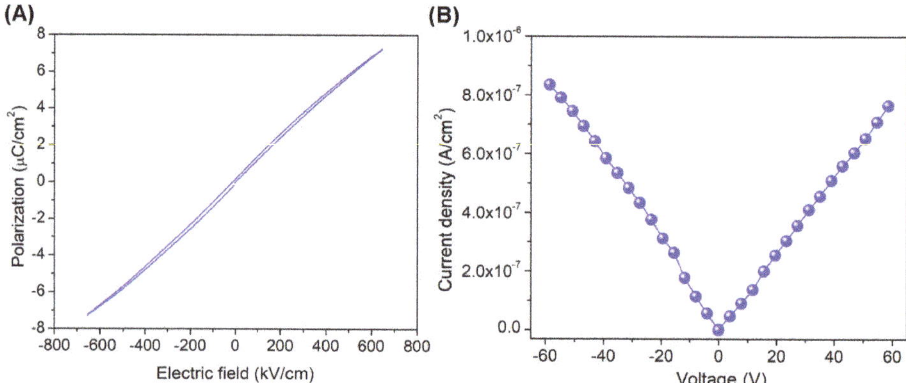

**Figure 3.** (**A**) The polarization-electric field (*P-E*) hysteresis loops measured at 100 Hz. (**B**) Leakage current density of the BSZT thin film capacitor as a function of the applied bias voltage at room temperature.

The electrical performance of a dielectric capacitor is determined mainly by the electric breakdown field ($E_b$) and dielectric constants ($\varepsilon_r$) of the dielectric layers. It is very difficult to obtain a dielectric material with the simultaneous features of high $\varepsilon_r$ and $E_b$ since there is an inherent tradeoff between the $\varepsilon_r$ and the $E_b$ for a dielectric material [2]. There is an equation used to describe the relationship between $\varepsilon_r$ and $E_b$:

$$\varepsilon_r E_b^2 = \text{Const} \tag{1}$$

This experimental constant is given as 400 when $E_b$ is expressed in megavolts per centimeter (MV/cm). The $\varepsilon_r E_b^2$ of most dielectric films is not more than 400. Figure 4 shows the $E_b$ as a function of $\varepsilon_r$ for a variety of dielectric thin-films [2,22,23]. In this work, the $E_b$ of the BSZT thin film was about 4 MV/cm, which is higher than the reported STO and BST films. Therefore, $\varepsilon_r E_b^2$ is higher than the constant. This result shows that the BSZT films are a very potential dielectric material for the thin filmcapacitor application in many performance parameters.

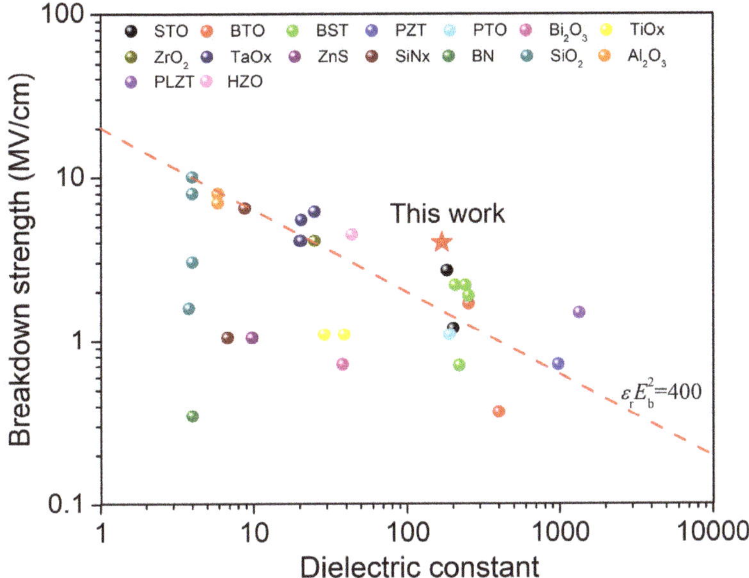


**Figure 4.** Electric breakdown field ($E_b$) as a function of dielectric constants ($\varepsilon_r$) for a variety of the dielectric thin-films. The data are from previously published literature.

## 4. Conclusions

In this work, high-quality BSZT thin films were successfully prepared on a Si/SiO$_2$/Pt(111) substrate. The prepared perovskite BSZT film also showed a high capacitance density of 1.7 nF/mm$^2$ at 100 kHz and a very low dielectric loss of 0.01 at 100 kHz. The prepared BSZT thin films also display a very low leakage current density (7.65 × 10$^{-7}$ A/cm$^2$ at 60 V) and high electric breakdown strength (~4 MV/cm). The results indicate that the BSZT thin film shows great potential for coupling and decoupling thin filmcapacitor applications.

**Author Contributions:** Conceptualization, P.Y.; methodology, X.C.; investigation, X.C., T.M., B.H., and Y.L.; data curation, X.C.; writing—original draft preparation, X.C.; writing—review and editing, P.Y.; project administration, P.Y.; funding acquisition, P.Y. All authors have read and agreed to the published version of the manuscript.

**Funding:** This research was funded by the National Natural Science Foundation of China under grant No. u1601208 and No. 51802204, China.

**Acknowledgments:** We appreciate Wang Hui from the Analytical and Testing Center of Sichuan University for her help with SEM characterization.

**Conflicts of Interest:** The authors declare no conflicts of interest.

## References


1. Chen, X.; Zhang, Y.; Xie, B.; Huang, K.; Wang, Z.; Yu, P. Thickness-Dependence of growth rate, dielectric response, and capacitance properties in $Ba_{0.67}Sr_{0.33}TiO_3/LaNiO_3$ hetero-structure thin films for film capacitor applications. *Thin Solid Films* **2019**, *685*, 269–274. [CrossRef]
2. Johari, H.; Ayazi, F. High-Density Embedded Deep Trench Capacitors in Silicon with Enhanced Breakdown Voltage. *IEEE Trans. Compon. Packag.* **2010**, *32*, 808–815. [CrossRef]
3. Perng, T.H.; Chien, C.H.; Chen, C.W.; Lehnen, P.; Chang, C.Y. High-Density MIM capacitors with $HfO_2$ dielectrics. *Thin Solid Films* **2004**, *469*, 345–349. [CrossRef]
4. Tuichai, W.; Danwittayakul, S.; Maensiri, S.; Thongbai, P. Investigation on temperature stability performance of giant permittivity (In + Nb) in co-Doped $TiO_2$ ceramic: a crucial aspect for practical electronic applications. *Rsc. Adv.* **2016**, *6*, 5582–5589. [CrossRef]
5. Pan, M.-J.; Randall, C.A. A brief introduction to ceramic capacitors. *IEEE Electr. Insul. Mag.* **2010**, *26*, 44–50. [CrossRef]
6. Sugii, N.; Yamada, H.; Kagaya, O.; Yamasaki, M.; Sekine, K.; Yamashita, K.; Watanabe, M.; Murakami, S. High-Frequency properties of $SrTiO_3$ thin-film capacitors fabricated on polymer-Coated alloy substrates. *Appl. Phys. Lett.* **1998**, *72*, 261–263. [CrossRef]
7. Kozyrev, A.; Samoilova, T.; Golovkov, A.; Hollmann, E.; Kalinikos, D.; Loginov, V.; Prudan, A.; Soldatenkov, O.; Galt, D.; Mueller, C. Nonlinear behavior of thin film $SrTiO_3$ capacitors at microwave frequencies. *J. Appl. Phys.* **1998**, *84*, 3326–3332. [CrossRef]
8. Finstrom, N.H.; Gannon, J.A.; Pervez, N.K.; York, R.A.; Stemmer, S. Dielectric losses of $SrTiO_3$ thin film capacitors with Pt bottom electrodes at frequencies up to 1 GHz. *Appl. Phys. Lett.* **2006**, *89*, 242910. [CrossRef]
9. Konofaos, N.; Evangelou, E.; Wang, Z.; Helmersson, U. Properties of $Al$-$SrTiO_3$-ITO capacitors for microelectronic device applications. *IEEE Trans. Electron Devices* **2004**, *51*, 1202–1204. [CrossRef]
10. Takemura, K.; Kikuchi, K.; Ueda, C.; Baba, K.; Aoyagi, M.; Otsuka, K. SrTiO3 thin film decoupling capacitors on Si interposers for 3D system integration. In Proceedings of the 2009 IEEE International Conference on 3D System Integration, San Francisco, CA, USA, 28–30 September 2009; pp. 1–5.
11. Takemura, K.; Ohuchi, A.; Shibuya, A. Si interposers integrated with $SrTiO_3$ thin film decoupling capacitors and through-Si-vias. In Proceedings of the 2008 IEEE 9th VLSI Packaging Workshop of Japan, Kyoto, Japan, 1–2 December 2008; pp. 127–130.
12. Jain, P.; Rymaszewski, E.J. Embedded thin film capacitors-Theoretical limits. *IEEE Trans. Adv. Packag.* **2002**, *25*, 454–458. [CrossRef]
13. Chan, N.; Wang, D.; Wang, Y.; Dai, J.; Chan, H. The structural and in-Plane dielectric/ferroelectric properties of the epitaxial $(Ba, Sr)(Zr, Ti)O_3$ thin films. *J. Appl. Phys.* **2014**, *115*, 234102. [CrossRef]
14. Fu, S.L.; Ho, I.C.; Chen, L.S. Studies on semiconductive $(Ba_{0.8}Sr_{0.2})(Ti_{0.9}Zr_{0.1})O_3$ ceramics. *J. Mater. Sci.* **1990**, *25*, 4042–4046. [CrossRef]
15. Jiang, L.; Tang, X.; Jiang, Y.; Liu, Q.; Ma, C.; Chan, H. Epitaxial growth and dielectric properties of BSZT thin films on $SrTiO_3$: Nb single crystal substrate prepared by pulsed laser deposition. *Surf. Coat. Technol.* **2013**, *229*, 162–164. [CrossRef]
16. Fan, Y.; Yu, S.; Sun, R.; Li, L.; Yin, Y.; Du, R. Microstructure and electrical properties of $Ba_{0.7}Sr_{0.3}(Ti_{1-x}Zr_x)O_3$ thin films prepared on copper foils with sol–gel method. *Thin Solid Films* **2010**, *518*, 3610–3614. [CrossRef]
17. Sharma, S.; Ram, M.; Thakur, S.; Sharma, H.; Negi, N.S. Structural Studies of Zirconium Doped $Ba_{0.70}Sr_{0.30}TiO_3$ Lead Free Ferroelectric Thin Films. *Aip Conf. Proc.* **2016**, *1728*, 020192.
18. Zhai, J.; Yao, X.I.; Chen, H. Structural and dielectric properties of $(Ba_{0.85}Sr_{0.15})(Zr_{0.18}Ti_{0.82})O_3$ thin films grown by a sol–gel process. *Ceram. Int.* **2004**, *30*, 1237–12403. [CrossRef]
19. Tang, X.; Jie, W.; Chan, L.W. Dielectric properties of columnar-grained $(Ba_{0.75}Sr_{0.25})(Zr_{0.25}Ti_{0.75})O_3$ thin films prepared by pulsed laser deposition. *J. Cryst. Growth* **2005**, *276*, 453–457. [CrossRef]
20. Chen, X.; Yang, C.; Zhang, X.; Li, J.; Liu, H.; Xiao, D.; Yu, P. The Effects of Buffers on the Microstructure and Electrical Properties of Mn/Y Co-Doped $Ba_{0.67}Sr_{0.33}TiO_3$ Thin Films. *Integr. Ferroelectr.* **2012**, *140*, 132–139. [CrossRef]
21. ABC of CLR in European Passive Components Institute (EPCI) Home Page. Available online: https://epci.eu/category/abc-of-clr/ (accessed on 26 March 2020).

22. Park, M.H.; Kim, H.J.; Kim, Y.J.; Moon, T.; Kim, K.D.; Hwang, C.S. Thin $Hf_xZr_{1-x}O_2$ films: a new lead-Free system for electrostatic supercapacitors with large energy storage density and robust thermal stability. *Adv. Energy Mater.* **2014**, *4*, 1400610. [CrossRef]
23. Ma, B.; Hu, Z.; Koritala, R.E.; Lee, T.H.; Dorris, S.E.; Balachandran, U. PLZT film capacitors for power electronics and energy storage applications. *J. Mater. Sci. Mater. Electron.* **2015**, *26*, 9279–9287. [CrossRef]

*Article*

# Enhanced Electrocaloric Effect in 0.73Pb(Mg$_{1/3}$Nb$_{2/3}$)O$_3$-0.27PbTiO$_3$ Single Crystals via Direct Measurement

Biao Lu [1], Xiaodong Jian [1,2], Xiongwei Lin [1,2], Yingbang Yao [1,2], Tao Tao [1,2], Bo Liang [1,2], Haosu Luo [3] and Sheng-Guo Lu [1,2,*]

[1]  Guangdong Provincial Research Center on Smart Materials and Energy Conversion Devices, Guangdong Provincial Key Laboratory of Functional Soft Condensed Matter, School of Materials and Energy, Guangdong University of Technology, Guangzhou 510006, China; mse377@163.com (B.L.); jianxiaodong_gdut@126.com (X.J.); linxw_gdut_edu@126.com (X.L.); ybyao@gdut.edu.cn (Y.Y.); taotao@gdut.edu.cn (T.T.); liangbo@gdut.edu.cn (B.L.)

[2]  Dongguan South China Design Innovation Institute, Building D-1, University Innovation City Area, Songshan Lake, Dongguan 523808, China

[3]  The State Key Laboratory of High Performance Ceramics and Superfine Microstructure, Shanghai Institute of Ceramics, Chinese Academy of Sciences, 215 Chengbei Road, Jiading, Shanghai 201800, China; hsluo@mail.sic.ac.cn

*  Correspondence: sglu@gdut.edu.cn

Received: 31 March 2020; Accepted: 27 May 2020; Published: 31 May 2020

**Abstract:** Electrocaloric properties of [110] and [111] oriented 0.73Pb(Mg$_{1/3}$Nb$_{2/3}$)O$_3$-0.27PbTiO$_3$ single crystals were studied in the temperature range of 293–423 K. The Maxwell relations and the Landau–Ginsburg–Devonshire (LGD) phenomenological theory were employed as the indirect method to calculate the electrocaloric properties, while a high-resolution calorimeter was used to measure the adiabatic temperature change of the electrocaloric effect (ECE) directly. The results indicate that the directly measured temperature changes of $\Delta T > 2.5$ K at room temperature were procured when the applied electric field was reversed from 1 MV/m to −1 MV/m, which are larger than those deduced pursuant to the Maxwell relation, and even larger than those calculated using the LGD theory in the temperature range of 293–~380 K.

**Keywords:** PMN-PT; single crystals; P–E hysteresis loop; electrocaloric effect; Maxwell relation

---

## 1. Introduction

The electrocaloric effect (ECE) is the adiabatic temperature change resulting from the polarization change in a polar material upon the application or removal of an electric field [1]. Theoretical studies indicate that cooling devices based on the ECE have a much higher energy conversion efficiency (>60% of the Carnot efficiency) than those of a vapor compressor [2]. The cooling technology based on the ECE may lead to more efficient and environmentally friendly alternative cooling technology. Due to the ECE caused by polarization change, strongly correlated polar materials, e.g., ferroelectrics and antiferroelectrics, will be the promising ECE candidates [3,4]. These materials will offer the potential to be applied in solid-state refrigeration.

In general, the ECE can be measured using a direct or an indirect approach. For the indirect one, although recently a few methodologies have been proposed to measure the ECE, the most convenient way is still the calculation using the Maxwell relation. Many researchers have adopted the phenomenological approach and a number of works on the ECEs in various materials were reported. The reason why the indirect method has become so widely accepted and popular is probably due to the fact that it is hard to measure the tiny ECEs (because of very small heat generated) using a direct way for thin films or thick films (despite their ECE temperature changes being relatively large due to the higher electric field applied). In addition, it is quite easy to use the probe station to measure the polarization for the thin/thick films, which thus makes the indirect method rather popular [5]. The phenomenological theories, however, are suitable for ideal situations, e.g., ergodic systems, single domains, etc. [1]. Hence, in practice, the ECE and other properties deduced using the phenomenological theory are not consistent with the experimental results because of the relaxation in the ferroelectrics and difficulty in forming the single domain. To address this issue, a few direct measurement setups have actually been designed and built. For the direct method, a convenient way which utilizes a thermometer or a thermistor directly attached to the sample seems to be a reliable fashion to obtain the ECE. Although this method is not carried out in an adiabatic condition, the temperature change can be measured before any significant heat exchanging with the surrounding can take place when the voltage is ramped up or down fast enough, e.g., in milliseconds. Meanwhile, the test data of this method can be further improved to ensure the measuring accuracy [5].

As important relaxor ferroelectrics, $(1-x)Pb(Mg_{1/3}Nb_{2/3})O_3$-$xPbTiO_3$ (PMN-PT) single crystals have been studied extensively [6–9]. The interesting properties of PMN-PT crystals are their high permittivity, diffuse phase transition, high piezoelectric constant and large electromechanical coupling factor. It is well known that PMN-PT crystals have been widely used as piezoelectric materials in industry and daily life. Moreover, PMN-PT crystals possess excellent polarization properties thus it is worth studying their electrocaloric properties. In this work, the ECEs of $[110]$-$0.73Pb(Mg_{1/3}Nb_{2/3})O_3$-$0.27PbTiO_3$ (0.73PMN–0.27PT) and $[111]$-0.73PMN–0.27PT crystals were calculated in accordance with the Maxwell relation and Landau–Ginsburg–Devonshire (LGD) phenomenological theory. In addition, a high-resolution calorimeter was employed to make accurate measurements of the temperature change due to the ECE induced by a change in the applied electrical field. The ECEs obtained by different approaches were compared and the discrepancies were also discussed.

## 2. Materials and Methods

The 0.73PMN–0.27PT single crystals were grown by making use of a modified Bridgman method. The obtained PMN-PT crystals were sliced into 0.5 mm-thick (110) and (111) plates. The surfaces of the PMN-PT plates were polished carefully and both surfaces were covered with gold as contact electrodes for the prospective test.

The permittivities as a function of temperature and frequency were measured using a precision impedance analyzer (Agilent 4284A, Agilent Technologies Inc., Santa Clara, CA, USA) equipped with a programmed temperature controller at 1 V and zero-bias field. The experimental data were acquired at a heating run at 1, 10, and 100 kHz. The specific heat capacities were obtained using a differential scanning calorimeter (DSC, Mettler-Toledo DSC-3, Mettler-Toledo (Schweiz) GmbH, Greifensee, Switzerland) in a modulated mode. The polarization–electric field (P–E) hysteresis loop was procured by a Sawyer–Tower circuit equipped with a temperature chamber (Radiant Multiferroic System, Radiant Technologies Inc., Albuquerque, NM, USA). The ECE measurement was pursued within the temperature range of 293 and 423 K with an increment of 10 K [10].

For the direct ECE measurement, a high-resolution thermometer (Omega T type thermocouple, Omega Engineering Inc., Norwalk, CT, USA) was employed and attached to the sample surface tightly. A high-voltage supply (Trek, 610E, Trek, Pennington, MN, USA) was used to generate the electric signal, which was then amplified and applied to the crystal sample. In the course of the experiment, a ramped voltage was exerted to the sample, and the electric field was kept at a constant value for over 10 s, in order to reach a thermal equilibrium with the surrounding [10]. After that, the voltage was sharply removed. The typical thermal response times along the sample thickness direction is a few milliseconds [11]. A thermal equilibrium throughout the whole sample, including the electrodes, attached thermometer and wires, can be reached within a short time. However, the time for the equilibrated crystal to exchange heat with the surrounding bath ($T_{bath}$) will take a longer time. The relaxation of the temperature of the whole system can be illustrated as follows [12],

$$T(t) = T_{bath} + \Delta T e^{-\frac{t}{\tau}} \tag{1}$$

Here, $\tau$ is the relaxation time constant. More details referring to the test details and data analysis can be obtained in Ref. [11]. In doing the test, the crystal was immersed in the silicone oil, which was then placed inside a temperature chamber (Delta Chamber, Delta 9064, Delta Design Inc., Poway, CA, USA). The chamber's temperature can be controlled using a computer and the temperature resolution for each test point is below ±0.1 °C. In the course of measurement, to mitigate the impact of the thermocouple, glue, and wires on the measuring accuracy, very slim silver wires (0.05 mm in diameter), a thermocouple with a tiny bead (0.1 mm in diameter) with very slim wires (0.05 mm in diameter), and minimal amounts of electrical and thermal conductive adhesives were used during the preparation of samples. Because of the very tiny contact areas between the sample and the wires, and very short testing time, the heat dissipated by wires is estimated to be quite small compared to the heat generated by the ECE crystal. Thus, Equation (1) was just used to correct the measured results [10].

## 3. Results

In the first indirect method, according to the well-known Maxwell relation $\left(\frac{\partial P}{\partial T}\right)_E = \left(\frac{\partial S}{\partial E}\right)_T$, the ECE can be determined from the measurements of the temperature dependences of polarization ($P$) at a constant electric field ($E$), then the ECE can be calculated using the equation [13]

$$\Delta T = -\int_{E_1}^{E_2} \frac{T}{\rho C_E}\left(\frac{\partial P}{\partial T}\right)_E dE, \tag{2}$$

where $\rho$ is the material density and $C_E$ the specific heat capacity. The $E_1$ and $E_2$ denote the start and end electric fields applied. To use this indirect measurement, the necessary treatment process is as follows. Firstly, the relationships between the polarization and the electric field (hysteresis loops) at different temperatures should be tested over a certain temperature range. Secondly, the temperature dependence of polarization with the applied electric field can be obtained from the upper branches of the hysteresis loops for $E > 0$. The heat capacity and bulk density are also needed and can be measured using a DSC and densimeter. Then the ECE can be calculated.

In the second indirect method, the ECE can also be deduced by expanding the free energy in a power series of polarization. Based on the LGD phenomenological theory as used in the $Pb(ZrTi)O_3$ (PZT) system [14], the Gibbs free energy under a zero-stress condition for [110] and [111] polarization orientation can be illustrated as [15]:

$$\text{For the [110] direction, } \Delta G = 2\sigma_1 p_3^2 + (2\sigma_{11} + \sigma_{12})p_3^4 + 2(\sigma_{111} + \sigma_{112})p_3^6 \tag{3}$$

$$\text{For the [111] direction, } \Delta G = 3\sigma_1 p_3^2 + 3(\sigma_{11} + \sigma_{12})p_3^4 + 3(\sigma_{111} + \sigma_{112} + \sigma_{123})p_3^6 \tag{4}$$

where $p_i$, $\sigma_1$, $\sigma_{ij}$, and $\sigma_{ijk}$ ($i, j, k = 1, 2, 3$) are the polarization components, and dielectric stiffnesses at a constant stress, respectively. All of the dielectric stiffness coefficients are independent of temperature, except the parameter $\sigma_1$, which can be expressed as a linear temperature dependence based on the Curie–Weiss law [16]

$$\sigma_1 = \frac{T - T_0}{\varepsilon_0 C} \tag{5}$$

$C$ is the Curie constant, $\varepsilon_0$ the vacuum dielectric constant and $T_0$ the Curie–Weiss temperature. Usually, $C$ and $T_0$ can be obtained by fitting the inverse of the dielectric constant as a function of temperature in the paraelectric phase using the Curie–Weiss law [17]. In order to deduce the ECE based on the LGD theory, the temperature dependence of permittivity should be tested firstly to obtain the Curie constant. Then using the relations: $\frac{\partial \Delta G}{\partial T} = -\Delta S$ and $\Delta T = -\frac{T}{C_E}\Delta S$, then the ECE can be obtained.

Furthermore, due to the ECE causing the ferroelectric sample temperature change subject to the application or removal of the electric field, the temperature change can be detected by a high-resolution calorimeter. The details are mentioned in the Materials and Methods.

In this study, we applied the electric field of 1 MV/m on the sample and stayed for a few seconds to reach the temperature equilibrium first, then the electric field was removed immediately. Meanwhile, the first ECE signal appeared. After the temperature equilibration, a reversed electric field of 1 MV/m (−1 MV/m) was applied on the sample. Then the second ECE signal can be detected. Furthermore, during the same test run, it can also be noticed that the intensity of the first ECE signal is much smaller than that of the second ECE signal at low temperature but the intensity difference of the two ECE signals became narrowed with the increasing temperature. Figure 1A,D–F illustrates the ECE signals of [110] direction at 296 K, 353 K, 383 K and 418 K, respectively. The ECE signals of the [111] direction of PMN-PT single crystals demonstrate the same tendency as well and are not present here. Due to the electric breakdown, the measurements of the [110] direction sample above 373 K were not carried out. In general, before the fitting, by making use of the transformation of coordinate translation, peak value was removed to the zero, as described in Figure 1B,C. The red curves are the fitted curves using Equation (1). $\Delta T_{ECE}$ was obtained by extrapolating the fitting toward the time of the fall of the step-like pulse. In the direct measurement of $\Delta T$, one concern is the Joule heating in the samples, which will cause lowering of temperatures when the field is removed. However, during the test, the base line temperature $T-T_{bath}$ in Figure 1 is the same except for the application of or withdrawing of the electric field, which indicates that the observed temperature change is indeed due to the ECE from the crystals.

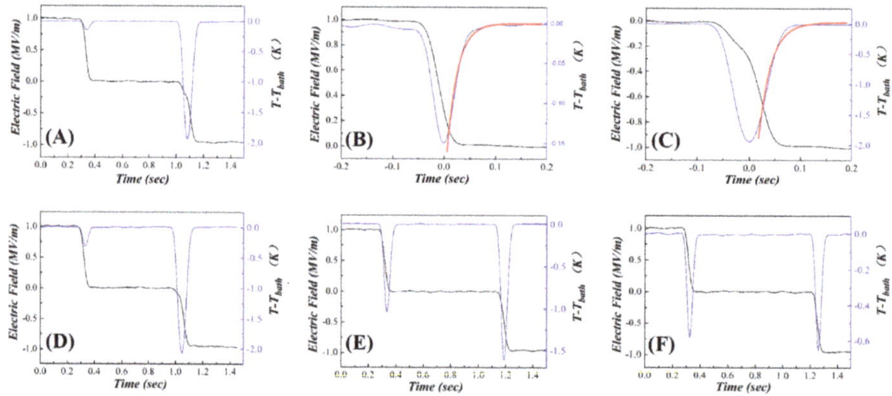

**Figure 1.** The electrocaloric effect (ECE) signals of [110] direction of PMN-PT single crystal at 1 MV/m. (**A**) ECE measured @ 296 K; (**B**) Fitting of 1st signal in Figure 1A; (**C**) Fitting of 2nd signal in Figure 1A; (**D–F**) ECE measured @ 353, 383, and 418 K, respectively.

The temperature dependences of permittivities are depicted in Figure 2A,B. The permittivities versus temperature at three frequencies (1, 10 and 100 kHz) exhibit board peaks in the measured temperature range. The P–E hysteresis loops were measured at 10 Hz and at a 10 K interval in the temperature range of 293 K to 423 K and the representative plots of P–E loops are shown in Figure 2C,D.

At low temperatures, the shape of the loops is rather square due to the absence of grain boundaries in single crystals [18]. With the increasing temperature, the hysteresis loops for the two samples exhibit a transition from square loops to slim ones. At each temperature, the remnant polarization ($P_r$) and spontaneous polarization ($P_s$) along the [110] direction are smaller than that of the [111] direction, but the coercive field ($E_c$) of [110] direction is slightly larger. There are eight equivalent polarization orientations along the [111] direction for rhombohedral relaxor ferroelectric PMN-27PT single crystals. When the electric field is applied on the sample, the dipoles will reorient along the electric field direction. While for the [110] direction, there are only two equivalent polar vectors along the [111] direction, which will rotate 35.5° toward the applied field direction of [110] with a designated domain engineered configuration "2R." [19]. For this case, the macroscopic symmetry is mm$^2$. In contrast, there is only one polar vector along the [111] direction for [111] poled ferroelectric crystals, thus it will form a monodomain state, designated as "1R", exhibiting symmetry 3 m [19]. According to the domain engineered configurations, the polarization level derived from the hysteresis loops should be correlated [20].

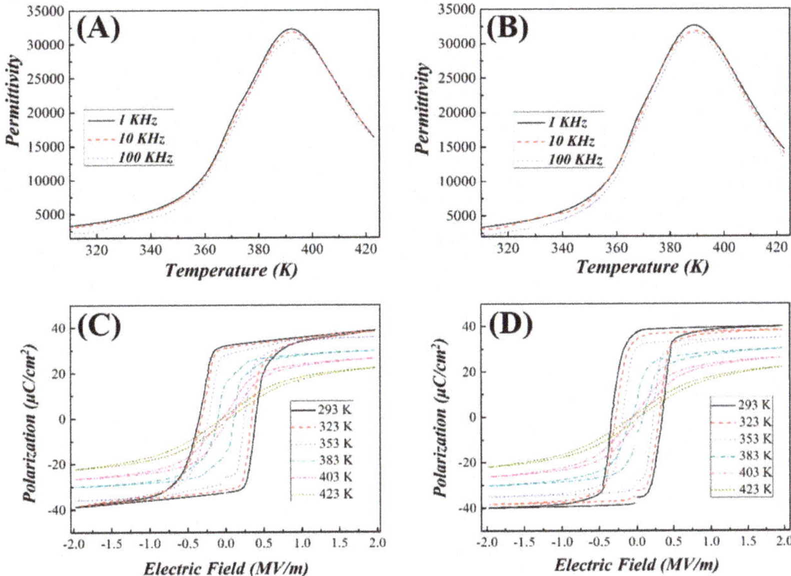

**Figure 2.** Permittivity and polarization–electric fields (P–E) hysteresis loop as a function of temperature for PMN-PT single crystals with the [110] direction (**A,C**) and [111] direction (**B,D**).

The ECE results obtained by indirect and direct ways are plotted in Figure 3. For the indirect method, the ECEs at 1 MV/m and 2 MV/m are present in the figures. For the ECE obtained indirectly, at the same electric field, the ECE calculated by the LGD theory is larger than that deduced by the Maxwell relation at any temperature. Moreover, the ECEs deduced via the Maxwell relation have peak values around the Curie temperature, about 400 K. The appearance of peak values is ascribed to the electric field induced phase transition from ferroelectric to paraelectric phase [21]. However, the ECEs calculated by the LGD theory decrease monotonically, even around the Curie temperature. For the ECEs obtained directly, the data measured at 1 MV/m and −1 MV/m are plotted respectively. Over the full testing temperature range, the measured data for the [110] direction sample at 1 MV/m are smaller below 383 K, but larger above 383 K than those calculated by the LGD theory, while the measured data of this sample are larger than those deduced by the Maxwell relation. For the [111] direction sample, due to the absence of the data at $\Delta T > 373$ K, the existing $\Delta T$ are between the data obtained by the two indirect methods. For the data measured at −1 MV/m below 383 K, the results are much larger than and in sharp contrast to those deduced from the indirect methods. The $\Delta T$ of the [110] direction sample at −1 MV/m and 1 MV/m have almost the opposite variation tendency. The $\Delta T$ of this sample at −1 MV/m is much larger than those at 1 MV/m below 393 K, but almost the same above 393 K. We noted the theoretical calculation using thermodynamic theory for PMN-0.35PT [22], for the electric field without causing flipping of the electric domains, our results are consistent with the calculation.

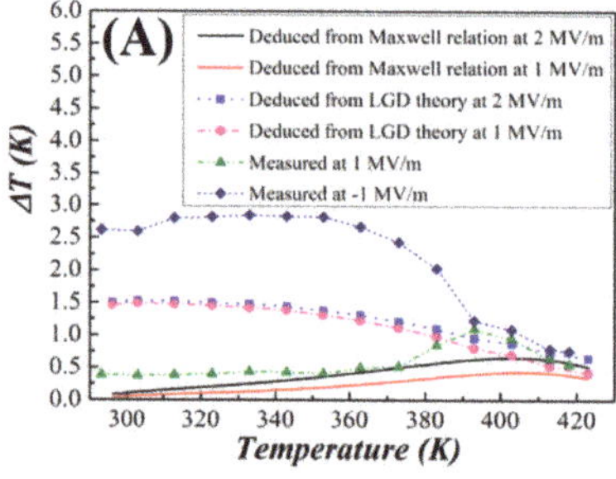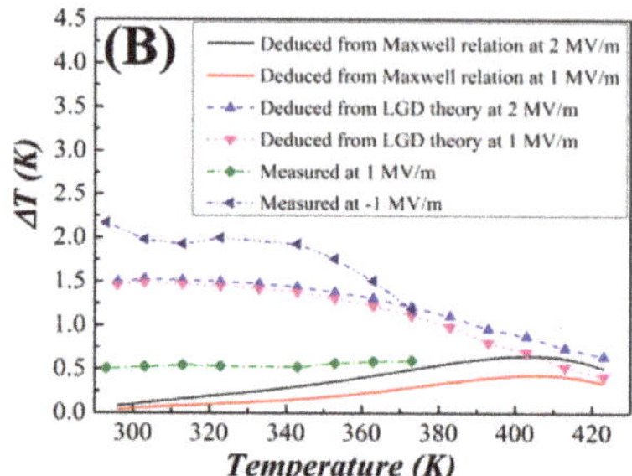

**Figure 3.** Temperature changes ($\Delta T$) for the [110] direction (**A**) and [111] direction (**B**) obtained from direct measurements, the Maxwell relation, and the LGD theory.

It can be seen from the P–E hysteresis loops (Figure 2), that below the temperature of 383 K, the shape of the P–E loops for 0.73PMN–0.27PT crystals is rather square. In that case, the values of the remnant polarization ($P_r$) are only slightly smaller than that of saturation polarization ($P_s$). Thus, when the electric field is released, the polarization variation is very small. That is, the change in the ordering degree of dipoles is very small. Hence, the polarization entropy increases little, while the lattice vibration entropy slightly decreases under the adiabatic condition. Hence, the ECE caused by the polarization entropy variation (the first ECE signal) is not significant. During the same test run, when the applied electric field changes to −1 MV/m, the dipoles are flip-switched and orient along the reversed electric field. During this process, the dipoles have a remarkable rotation and a large polarization entropy variation comes up. The distinct appearance of the second ECE signal is attributed to this process. The phenomena can be illustrated as follows. For the electric field changes from +E to 0, the polarization changes from $P_s$ to $P_r$, then the entropy change

$$\Delta S_1 = -\frac{1}{2}\frac{1}{\varepsilon_0 C}(P_r^2 - P_S^2) \tag{6}$$

For the electric field changes from 0 to −E, Basso et al. observed an enhancement of ECE in P(VDF-TrFE) polymers [23], but no illustration was reported. Considering the ferroelectric system as a quasi-one dimensional harmonic oscillator, the free energy can be expressed in terms of the harmonic term plus higher order terms, in which the harmonic term is proportional to the square of the average value of the canonical coordinate of the B-site ion, i.e., $\langle Q_1 \rangle$, which is corresponding to the order parameter-polarization in the Landau average field theory. For a displacive type ferroelectric, the polarization is proportional to the displacement of the ions. This means that the ionic displacement is proportional to the polarization. Now one can see that the ionic displacement changes from plus to minus when the electric field changes its sign, leading to the polarization changes from +P to −P. If the polarization changes from $P_r$ to −$P_s$, the ionic displacement changes from ~+$\langle Q_1 \rangle$ to −$\langle Q_1 \rangle$, totaled ~2$\langle Q_1 \rangle$. Thus, the harmonic energy has a change of about four times of the original value if the change of displacement is directly from +$P_r$ (~+$P_s$) to (−$P_s$). Thus, the isothermal entropy change will have a large enhancement compared with that when the electric field changes from E to 0. Therefore, a larger peak will be observed (shown in Figure 1). However, as shown in Figure 2, the P–E hysteresis loops have a transition from a square-like form to a slim form with the increasing temperature. The $P_r$ also becomes smaller and smaller with the increasing temperature. Then the entropy change in Equation (6) gets larger and leads to a large ECE signal. When the applied electric field was reversed to −1 MV/m, the displacement of B-site ion decreases, thus the second ECE signal becomes smaller as well. Therefore, as the temperature increases, the first ECE signal becomes larger compared with the second ECE signal during the same test process.

The calculation using the Maxwell relation is a convenient way to procure the ECE, and has been widely employed in procuring the ECE for crystals or polycrystalline materials [24], but it should be noted that the application of the Maxwell relation should be limited to continuous phase transition, and beyond the range of non-ergodicity. For the ECE deduced using the Maxwell relation, the temperature dependence of polarization *P(T)* is usually obtained by fitting the measured polarizations extracted from the P–E hysteresis loops at different temperatures. For instance, in this study, the P–E loops were gained at a 10 K interval. Within the two adjacent testing temperatures, the polarization can only be predicated by fitting the discrete data. Thus, the *P vs T* relationship cannot be determined accurately and this leads to errors for $\left(\frac{\partial P}{\partial T}\right)_E$ [25]. Within a dynamic measurement of the ECE in an AC electric field, the finite polarization relaxation time has an impact on $\left(\frac{\partial P}{\partial T}\right)_E$ and leads to errors for the quasi-static model [5,26].

The dielectric behavior of relaxor or normal ferroelectrics can be determined in terms of a universal Curie–Weiss law [27]:

$$\frac{1}{\varepsilon} - \frac{1}{\varepsilon_m} = \frac{(T - T_m)^{\gamma}}{C'} \tag{7}$$

where $\varepsilon_m$ and $T_m$ are the maximum dielectric constant and the corresponding temperature observed in a wide temperature range around the phase transition point, $\varepsilon$ and $T$ are the dielectric constant and the corresponding temperature above the $T_m$, $C'$ is the Curie-like constant, and $\gamma$ is the exponent indicating the diffusiveness of the ferroelectric. Ideally, when $\gamma = 1$, the material is a normal ferroelectric, but when $\gamma = 2$, the ferroelectric is a relaxor ferroelectric. The relaxor ferroelectric behavior is increasing gradually with $\gamma$ when $\gamma$ is between 1 and 2. $\gamma$ can be worked out via fitting the logarithmic plot of the reciprocal permittivity $(\frac{1}{\varepsilon} - \frac{1}{\varepsilon_m})$ as a function of temperature $(T - T_m)$, measured at the same frequency. As shown in Figure 4A,B, the $\gamma$ of the reciprocal permittivity with respect to temperature was determined from the slope of the fitted straight line: $\gamma = 1.92$ for the [110] direction sample and 1.87 for the [111] direction sample. Thus, the two samples have strong relaxor ferroelectric behaviors. For the relaxor ferroelectrics, a glassy polarization mechanism was subsequently proposed with correlations between super-paraelectric moments leading to the development of effective non-ergodicity in a frozen state [28]. Then, for the relaxor ferroelectrics, the relaxation time of polarization also limited the accuracy of the ECE deduced by the Maxwell relation since the Maxwell relation is derived based on the assumption that the thermodynamic system is ergodic [29]. Typical non-ergodic systems make the systems with very low relaxation times. The measuring procedure itself changes the state of the material (the so-called memory effect), which influences the measurement results [30].

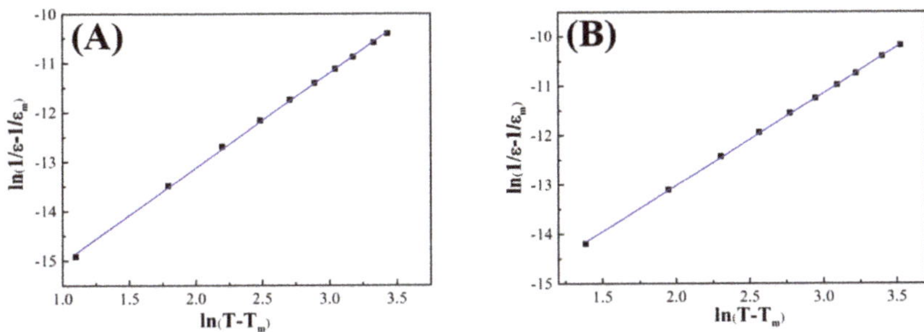

**Figure 4.** Reciprocal permittivity $(\frac{1}{\varepsilon} - \frac{1}{\varepsilon_m})$ (at 1 kHz) as a function of temperature $(T - T_m)$ for crystals with a direction of [110] (**A**), and [111] (**B**).

In a multi-domain material, the domain dynamics under the applied field reduce the excess entropy available for transfer to acoustic modes, consequently, reducing the ECE. To increase the accuracy of the indirect method, the system must be brought toward a single domain crystal, which can, in the first approximation, be achieved at the state of saturated polarization [31]. However, bulk crystal cannot sustain very high electric fields, so the single domain is difficult to form in the crystal.

As has been mentioned above, in the Gibbs free energy equation, the coefficient $\sigma_1$ was assumed approximately a linear temperature dependence as the usual expression for normal ferroelectric and $\sigma_1$ equals to $\frac{T-T_0}{\varepsilon_0 C}$. For normal ferroelectrics, the paraelectric phase appears instantly above the Curie temperature, but the ferroelectric phase still exists above the Curie temperature in relaxor ferroelectrics. According to Pirc's work [32], in relaxor ferroelectrics, the value of $\varepsilon_0 C$ is not constant and expected to be a function of temperature because the coefficient $\sigma_1$ must remain positive at all temperatures. Thus, large errors occur when deducing the Curie constant $C$. Moreover, the existence of domain walls, domain structure, domain wall mobility and defect dipoles will impact the polarization [33], and cause error to the ECEs deduced from the Gibbs free energy equation.

For the ECE measured directly, the main source of measurement error is due to the heat that dissipates through objects attached to the sample and surrounding around the sample, e.g., the thermocouple and the electrode wires. However, due to fast internal thermal response, the method used in this study has a sufficient accuracy [34]. Therefore, the ECE measured directly is considered to be the closest to the real situation.

## 4. Conclusions

In conclusion, a large ECE was procured through reversing the applied electric field, which is larger than that by applying the electric field in the same direction for 0.73PMN–0.27PT relaxor ferroelectric crystals. The directly measured results are larger than those calculated from the Maxwell relation and even larger than those deduced using the LGD theory in the lower temperature range. The flipping of electric domains can be used to further enhance the ECE in ferroelectrics.

**Author Contributions:** Conceptualization, S.-G.L. and B.L. (Biao Lu); methodology, B.L. (Biao Lu); software, X.L. and Y.Y.; validation, B.L. (Biao Lu); X.J., Y.Y., T.T., B.L. (Bo Liang), H.L. and S.-G.L.; formal analysis, B.L. (Biao Lu); investigation, B.L. (Biao Lu); sources, H.L.; writing—original draft preparation, B.L. (Biao Lu); writing—review and editing, S.-G.L.; supervision, S.-G.L.; funding acquisition, S.-G.L. All authors have read and agreed to the published version of the manuscript.

**Funding:** This research was funded by the Natural Science Foundation of China (Grant No. 51372042, 51872053), Guangdong Provincial Natural Science Foundation (2015A030308004), the NSFC-Guangdong Joint Fund (Grant No. U1501246) and the Dongguan Frontier Research Project (2019622101006).

**Conflicts of Interest:** The authors declare no conflict of interest.

## References

1. Lines, M.E.; Glass, A.M. *Principles and Applications of Ferroelectrics and Related Materials*; Clarendon Press: Oxford, UK, 1977.
2. Pakhomov, O.V.; Karmanenko, S.F.; Semenov, A.A.; Starkov, A.S.; Es'kov, A.V. Thermodynamic estimation of cooling efficiency using an electrocaloric solid-state line. *Tech. Phys.* **2010**, *55*, 1155–1160. [CrossRef]
3. Mischenko, A.S.; Zhang, Q.; Scott, J.F.; Whatmore, R.W.; Mathur, N.D. Giant electrocaloric effect in thin-film $PbZr_{0.95}Ti_{0.05}O_3$. *Science* **2006**, *311*, 1270–1271. [CrossRef] [PubMed]
4. Neese, B.; Chu, B.; Lu, S.G.; Wang, Y.; Furman, E.; Zhang, Q.M. Large electrocaloric effect in ferroelectric polymers near room temperature. *Science* **2008**, *321*, 821–823. [CrossRef] [PubMed]
5. Liu, Y.; Scott, J.F.; Dkhil, B. Direct and indirect measurements on electrocaloric effect: Recent developments and perspectives. *Appl. Phys. Rev.* **2016**, *3*, 031102. [CrossRef]
6. Ma, Y.B.; Novak, N.; Koruza, J.; Yang, T.Q.; Albe, K.; Xu, B.X. Enhanced electrocaloric cooling in ferroelectric single crystals by electric field reversal. *Phys. Rev. B* **2016**, *94*, 100104. [CrossRef]
7. Zhang, T.F.; Tang, X.G.; Ge, P.Z.; Liu, Q.X.; Jiang, Y.P. Orientation related electrocaloric effect and dielectric phase transitions of relaxor PMN-PT single crystals. *Ceram. Int.* **2017**, *43*, 16300–16305. [CrossRef]
8. Wu, H.H.; Cohen, R.E. Electric-field-induced phase transition and electrocaloric effect in PMN-PT. *Phys. Rev. B* **2017**, *96*, 054116. [CrossRef]
9. Sebald, G.; Seveyrat, L.; Guyomar, D.; Lebrun, L.; Guiffard, B.; Pruvost, S. Electrocaloric and pyroelectric properties of $0.75Pb(Mg_{1/3}Nb_{2/3})O_3$-$0.25PbTiO_3$ single crystals. *J. Appl. Phys.* **2006**, *100*, 124112. [CrossRef]

10. Lu, B.; Yao, Y.B.; Jian, X.D.; Tao, T.; Liang, B.; Zhang, Q.M.; Lu, S.G. Enhancement of the electrocaloric effect over a wide temperature range PLZT ceramics by doping with $Gd^{3+}$ and $Sn^{4+}$ ions. *J. Eur. Ceram. Soc.* **2019**, *39*, 1093–1102. [CrossRef]

11. Kutnjak, Z.; Rožič, B. Indirect and direct measurements of the electrocaloric effect. In *Electrocaloric Material*; Correia, T., Zhang, Q., Eds.; Springer: Berlin/Heidelberg, Germany, 2014; pp. 147–182.

12. Yao, H.; Ema, K.; Garland, C.W. Nonadiabatic scanning calorimeter. *Rev. Sci. Instrum.* **1998**, *69*, 172–178. [CrossRef]

13. Lu, S.G.; Zhang, Q.M. Electrocaloric materials for solid state refrigeration. *Adv. Mater.* **2009**, *21*, 1983–1987. [CrossRef]

14. Haun, M.J.; Furman, E.; McKinstry, H.A.; Cross, L.E. Thermodynamic theory of the lead zirconate-titanate solid solution system, part II: Tricritical behavior. *Ferroelectrics* **1989**, *99*, 27–44. [CrossRef]

15. Haun, M.J.; Furman, E.; Jang, S.J.; Cross, L.E. Thermodynamic theory of the lead zirconate-titanate solid solution system, part I: Phenomenology. *Ferroelectrics* **1989**, *99*, 13–25. [CrossRef]

16. Baerwald, H.G. Thermodynamic theory of ferroelectric ceramics. *Phys. Rev.* **1957**, *105*, 480–486. [CrossRef]

17. Lu, S.G.; Rozic, B.; Kutnjak, Z.; Zhang, Q.M. Electrocaloric effect in ferroelectric P(VDF-TrFE) copolymers. *Integr. Ferroelectr.* **2011**, *125*, 176–185. [CrossRef]

18. Marincel, D.M.; Zhang, H.R.; Kumar, A.; Jesse, S.; Kalinin, S.V.; Rainforth, W.M.; Reaney, I.M.; Randall, C.A.; Trolier-McKinstry, S. Influence of a single grain boundary on domain wall motion in ferroelectrics. *Adv. Funct. Mater.* **2014**, *24*, 1409–1417. [CrossRef]

19. Zhang, S.; Sherlock, N.P.; Meyer, R.J.; Shrout, T.R. Crystallographic dependence of loss in domain engineered relaxor-PT single crystals. *Appl. Phys. Lett.* **2009**, *94*, 162906. [CrossRef]

20. Davis, M.; Damjanovic, D.; Hayem, D.; Setter, N. Domain engineering of the transverse piezoelectric coefficient in perovskite ferroelectrics. *J. Appl. Phys.* **2005**, *98*, 014102. [CrossRef]

21. Shi, Y.P.; Soh, A.K. Modeling of enhanced electrocaloric effect above the Curie temperature in relaxor ferroelectrics. *Acta Mater.* **2011**, *59*, 5574–5583. [CrossRef]

22. Khassaf, H.; Mantese, J.V.; Bassiri-Gharb, N.; Kutnjak, Z.; Alpay, S.P. Perovskite ferroelectrics and relaxor-ferroelectric solid solutions with large intrinsic electrocaloric response over broad temperature ranges. *J. Mater. Chem. C* **2016**, *4*, 4763. [CrossRef]

23. Basso, V.; Gerard, J.-F.; Pruvost, S. Doubling the electrocaloric cooling of poled ferroelectric materials by bipolar cycling. *Appl. Phys. Lett.* **2014**, *105*, 052907. [CrossRef]

24. Lu, S.G.; Zhang, Q.M. Large electrocaloric effect in relaxor ferroelectrics. *J. Adv. Dielectr.* **2012**, *2*, 123011. [CrossRef]

25. Lu, B.; Li, P.L.; Tang, Z.H.; Yao, Y.B.; Gao, X.S.; Kleemann, W.; Lu, S.G. Large electrocaloric effect in relaxor ferroelectric and antiferroelectric lanthanum doped lead zirconate titanate ceramics. *Sci. Rep.* **2017**, *7*, 45335. [CrossRef] [PubMed]

26. Hanrahan, B.; Espinal, Y.; Neville, C.; Rudy, R.; Rivas, M.; Smith, A.; Kesim, M.T.; Alpay, S.P. Accounting for the various contributions to pyroelectricity in lead zirconate titanate thin films. *J. Appl. Phys.* **2018**, *123*, 124104. [CrossRef]

27. Bokov, A.A.; Ye, Z.G. Dielectric relaxation in relaxor ferroelectrics. *J. Adv. Dielectr.* **2012**, *2*, 1241010. [CrossRef]

28. Viehland, D.; Jang, S.J.; Cross, L.E.; Wuttig, M. Deviation from Curie-Weiss behavior in relaxor ferroelectrics. *Phys. Rev. B* **1992**, *46*, 8003. [CrossRef]

29. Lu, S.G.; Rožič, B.; Zhang, Q.M.; Kutnjak, Z.; Pirc, R.; Lin, M.R.; Li, X.Y.; Gorny, L.J. Comparison of directly and indirectly measured electrocaloric effect in the relaxor ferroelectric polymers. *Appl. Phys. Lett.* **2010**, *97*, 202901. [CrossRef]

30. Valant, M. Electrocaloric materials for future solid-state refrigeration technologies. *Prog. Mater. Sci.* **2012**, *57*, 980–1009. [CrossRef]

31. Birks, E.; Dunce, M.; Peräntie, J.; Hagberg, J.; Sternberg, A. Direct and indirect determination of electrocaloric effect in $Na_{0.5}Bi_{0.5}TiO_3$. *J. Appl. Phys.* **2017**, *121*, 224102. [CrossRef]

32. Pirc, R.; Kutnjak, Z.; Blinc, R.; Zhang, Q.M. Upper bounds on the electrocaloric effect in polar solids. *Appl. Phys. Lett.* **2011**, *98*, 021909. [CrossRef]

33. Weyland, F.; Bradesko, A.; Ma, Y.B.; Koruza, J.; Xu, B.X.; Albe, K.; Rojac, T.; Novak, N. Impact of polarization dynamics and charged defects on the electrocaloric response of ferroelectric $Pb(Zr,Ti)O_3$ ceramics. *Energy Technol.* **2018**, *6*, 1518–1525. [CrossRef]
34. Lu, S.G.; Rozic, B.; Zhang, Q.M.; Kutnjak, Z.; Li, X.Y.; Furman, E.; Gorny, L.J.; Lin, M.R.; Malic, B.; Kosec, M.; et al. Organic and inorganic relaxor ferroelectrics with giant electrocaloric effect. *Appl. Phys. Lett.* **2010**, *97*, 162904. [CrossRef]

MDPI

St. Alban-Anlage 66

4052 Basel

Switzerland

Tel. +41 61 683 77 34

Fax +41 61 302 89 18

www.mdpi.com

*Crystals* Editorial Office

E-mail: crystals@mdpi.com

www.mdpi.com/journal/crystals